P9-ELG-881

the great apes

the great apes

Jennifer Lindsey

FOREWORD BY JANE GOODALL

MetroBooks

MetroBooks
An Imprint of Friedman/Fairfax Publishers

Library of Congress Cataloging-in-Publication Data available upon request.

ISBN 1-56799-734-1

Editor: Susan Lauzau
Art Director: Jeff Batzli
Designer: Orit Mardkha-Tenzer
Photography Editor: Sarah Storey
Production Director: Karen Matsu Greenberg

Color separations by Spectrum Pte Ltd
Printed and bound in China by Leefung-Asco Printers Ltd.

1 3 5 6 9 10 8 6 4 2

For bulk purchases and special sales, please contact:
Friedman/Fairfax Publishers
Attention: Sales Department
15 West 26th Street
New York, NY 10010
212/685-6610 FAX 212/685-1307

Visit our website:
http://www.metrobooks.com

dedication

To Amahoro, Max, and Injii, the first great apes I ever met—For letting me into your world.

To Andy—For holding my hand as I wander.

To Brandy, Andrew, Kate, and Spencer—and to all my other nieces and nephews—Your generation holds the great apes' future in its hands.

acknowledgments

Had I never read Jane Goodall's book *In the Shadow of Man,* I would never have known the tragic, heartwarming, and fascinating lives of the chimpanzees of Gombe, never ventured to Africa to care for orphaned chimpanzees, nor devoted a career to their conservation and welfare. When I first opened *In the Shadow of Man,* I certainly could not have imagined that, years later, I would proudly call Jane Goodall my friend. Thank you, Jane, for your inspiration.

I gratefully acknowledge the dedication, persistence, and keen observations of the field researchers who have devoted their lives to studying and understanding the great apes—in the wild and in captivity. Particular thanks to Jane Goodall, Birute Galdikas, George Schaller, Frans de Waal, and the late Dian Fossey for their initial and ongoing forays into the mysterious worlds of these fascinating beings.

Thanks also to those who work tirelessly to save chimpanzees, bonobos, gorillas, and orangutans from extinction, including all those at the Jane Goodall Institute, the Dian Fossey Gorilla Fund, and Orangutan Foundation International. For assistance in the research of this book, I am very grateful to Eleanne van Vliet, who gave up many evenings and weekends in pursuit of the latest reports from the field. Many thanks to my editor, Susan Lauzau, for giving me the incredible opportunity to delve into a topic very close to my heart. And eternal thanks to my husband, Andy Nelson, who provides me with the support and encouragement to pursue my passion.

contents

foreword

by Jane Goodall

The forest was peaceful. Some 50 feet (15m) above me two chimpanzees were feeding—sixteen-year-old Prof and his brother, Pax, seven hears his junior. Since their mother's death five years before, Prof had looked after his little brother, waiting for him while traveling, searching and crying if they became separated, doing his best to protect his young charge when the need arose. Silhouetted against the sky, Prof swung to a new branch; the next moment I heard a crack and watched in horror as Prof hurtled toward the ground, desperately and uselessly grabbing at the vegetation. Fortunately, he fell onto a tangled mass of vines about 10 feet (3m) above the forest floor—and it held. But it had been a close call, and Prof sat motionless, shocked. In the silence I heard tiny whimpers, almost inaudible, and a rustling in the vegetation. Pax appeared. He climbed close to Prof and sat very still, staring, uttering tiny sounds of distress. Then, with an expression of extreme concern on his face, he moved right around his brother, examining him from every angle. Finally, as though satisfied that all was well, he began to groom Prof. And gradually Prof relaxed, soothed by those gently moving fingers.

It is nearly forty years since I began my chimpanzee research at Gombe—forty years during which I and my team have been privileged to live with and learn from these wonderful chimpanzees, compiling life histories, family histories, and the history of a community of complex beings who cannot record it themselves. Long-term studies of known groups of gorillas and orangutans started soon after mine, and finally research on bonobos began. As a result of books and documentaries about these great apes, thousands of people around the world have heard of Flo and Fifi and David Greybeard of Gombe. And Beethoven, Uncle Bert, and of course, Digit, the stars of Dian Fossey's study of mountain gorillas. Birute Galdikas describes the solitary lives of orangutans Princess and Supinah as they move through the Bornean rain forest.

Knowledge of these individuals has helped to shape our attitude toward the great apes and has blurred the line, once seen so sharp, between humans and the rest of the animal kingdom. For we know that the great apes have vivid personalities, are capable of rational thought and simple problem solving, and have emotions similar to those we call joy, sadness, anger, fear, and despair. We know they are capable of acts of altruism and have a concept of self. When I gaze inquiringly into the eyes of a chimpanzees, gently and with no trace of superiority, the chimpanzee will often return the look. There is a meeting of minds. For without doubt the chimpanzee has a mind—a mind that is far more similar to the human mind than anyone would have believed forty years ago. And the same is true for the other great ape species.

How sad that, just as we are beginning to understand these humanlike beings they are rapidly disappearing in the wild. When I began my research there were between one and two million chimpanzees living in twenty-five African nations. Today, 120,000 to 150,000, at most, remain, spread across twenty-one countries. Many of the remaining populations are living in tiny fragmented patches of forest surrounded by cultivated fields—and once the number of chimps in a given area drops below about 150, that group is probably doomed in the long run because they will become weakened by inbreeding. Sadly, this is true for the Gombe population. The national park, on the eastern shore of Lake Tanganyika, is only 30 square miles ($78km^2$) Within this tiny area there are some 120 individuals living in three communities—and they are totally isolated from all other chimpanzees. In 1960 chimpanzee habitat stretched northward from the park to the Burundi border and beyond, southward to Zambia and eastward as far as the eye could see. Today, outside the park, the forest has gone. The surrounding human population has mushroomed, needing ever more land for growing crops, grazing goats, and building houses. The soil has become infertile, desertlike, but the peasants are too poor to buy food. The same grim picture is repeated again and again in areas where the great apes once held sway. All are endangered.

There is still some hunting of the great apes for the live animal trade—hunters shoot ape mothers to steal their babies for pets, entertainment, or medical research. But today a major threat to chimpanzees, lowland gorillas, and bonobos is commercial hunting for food—in some regions it is a far more serious problem than habitat construction. Logging companies have carved roads deep in the last great forests, and hunters from the towns travel the logging roads and shoot everything they see—great apes, monkeys, okapi, elephants, and even small birds and bats. And the logging trucks take the meat to the big markets in the towns, where there is a cultural preference for meat from wild animals. People will actually pay more for it than the meat from domestic farm animals.

It was when I suddenly realized the terrible plight of chimpanzees across Africa that I knew I had to devote my energy to conservation. Conservation education programs, especially in the great ape range countries, are very important. It is necessary to find ways to improve the lives of the people living around protected areas of forest, so that they realize there is something in all this for them, so that they understand the need to protect the remaining forests. For without their cooperation, our efforts are doomed. Carefully controlled ecotourism can help as well.

We cannot save all the wild populations of great apes, but if we act together, each individual making contributions of time, talent, or money, we can save some of them. And we can improve the conditions of captive apes around the world. Fortunately, more and more people are prepared to act. One such individual is Jennifer Lindsey, author of this book and my coworker and friend. Jennifer is a writer with a passion to help our closest relatives. This book was written because of her deep concern and is her contribution to the cause. And it will make a difference. It will introduce the great apes to countless readers, and they will become involved; they will want to help. Because Jennifer writes with skill and clarity, from personal knowledge, and, most importantly, from her heart.

chapter i

introducing the great apes

Deep within the lush green vegetation of the Congo basin, high in the peat-swamp forests of Borneo, and nestled in the deciduous woodland of eastern Tanzania live humans' closest relatives in the animal world—the great apes. The chimpanzee, bonobo, orangutan, and gorilla are grouped with humans in the taxonomic order of primates, a classification created by the Swedish botanist Linnaeus in 1758. Linnaeus also included monkeys and lemurs in the primate order, acknowledging even then the striking physical similarities among these diverse species.

Great apes join humans in the taxonomic superfamily of Hominoidea, a higher classification of primates than monkeys. The physical differences between great apes and other nonhuman primates are obvious. Great apes have no tail, are larger than monkeys, and have a broader chest and more upright body posture. Great apes, like humans, have larger skulls that house bigger brains; they have short, broad noses rather than snouts; and they are less arboreal than the lower primates. But the traits that distinguish great apes from monkeys extend far beyond physical characteristics. The closest living relatives of humans, great apes mirror our own species in behavior and cognition in ways we are still discovering, and it is these attributes that divide them so definitely from the lower primates.

When Charles Darwin proposed boldly in 1859 that humans had evolved from apes, he pointed not only to the anatomical similarities; our minds, he said, are also the same. "There is no fundamental difference between man and the higher mammals in the mental faculties," Darwin proclaimed in his 1871 book *The Descent of Man*. Thanks to recent scientific research, we now know that Darwin's assertion regarding the physical resemblance between man and ape was astonishingly accurate. We share 98.4 percent of our DNA—our genetic makeup—with the chimpanzee and bonobo, making us more similar to these two species than foxes are to dogs. And the gorilla and orangutan come in close behind, with a difference from human DNA of 2.3 percent in gorillas and 4 percent in orangutans. The DNA of chimpanzees and Old World monkeys, by comparison, differs by 7.3 percent. In short, humans and great apes are more closely related to each other than either is to monkeys.

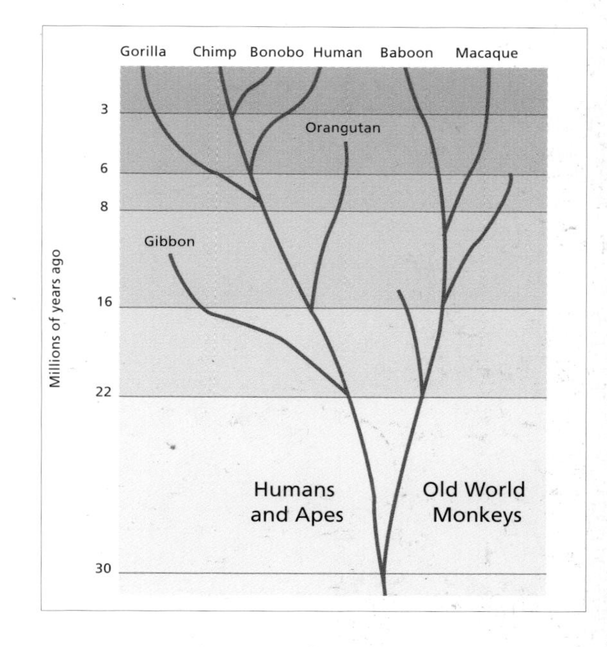

above: *This family tree shows the evolution of Old World primates. The model was developed using information gained from DNA.*

opposite: *Chimpanzees share more than 98.4 percent of their genetic makeup with humans, which means that they are closer biologically to humans than they are to monkeys. Although scientists and laypeople struggle with the implications of these statistics, few can deny the obvious physical similarities between humans and the great apes.*

the family of great apes

ORDER: PRIMATES
SUPERFAMILY: HOMINOIDEA
FAMILY: HOMINIDAE (chimpanzees, bonobos, gorillas—and humans)
FAMILY: PONGIDAE (orangutans)

common chimpanzee (*Pan troglodytes*)

three subspecies:

a. *Pan troglodytes troglodytes* (Central Africa)
b. *Pan troglodytes verus* (Western Africa)
c. *Pan troglodytes schweinfurthii* (Eastern Africa)

DISTRIBUTION: Across Equatorial Africa, from Guinea to western Tanzania and Uganda.
HABITAT: Tropical rain forest, savanna woodland, and montane forests.
POPULATION: 150–170,000
STATUS: Endangered, due to habitat destruction, poaching of infants, and hunting of adults for the bushmeat trade.

bonobo (*Pan paniscus*)

DISTRIBUTION: Congo (formerly Zaire), south of the Zaire River.
HABITAT: Humid rain forests in the flat river basin.
POPULATION: 10,000
STATUS: Endangered, due to habitat destruction, poaching of infants, and hunting of adults for the bushmeat trade.

orangutan (*Pongo pygmaeus*)

two subspecies:

a. *Pongo pygmaeus pygmaeus* (Borneo)
b. *Pongo pygmaeus abelii* (Sumatra)

DISTRIBUTION: Northern Sumatra and most of Borneo.
HABITAT: Tropical montane forest, peat-swamp forest, and tropical rain forest.
POPULATION: 20,000–27,000
STATUS: Endangered, due to habitat destruction and poaching.

gorilla (*Gorilla gorilla*)

three subspecies:

a. *Gorilla gorilla gorilla* (Western lowland)
b. *Gorilla gorilla graueri* (Eastern lowland)
c. *Gorilla gorilla beringei* (Mountain)

DISTRIBUTION: Discontinuous forest clumps in parts of Western and Central Africa.
HABITAT: Tropical secondary forest and temperate mountainous hillsides.
POPULATION: 100,000 western lowlands; 10,000 eastern lowlands; 650 mountain gorillas
STATUS: Endangered, due to habitat destruction, poaching of infants, and hunting of adults for the bushmeat trade.

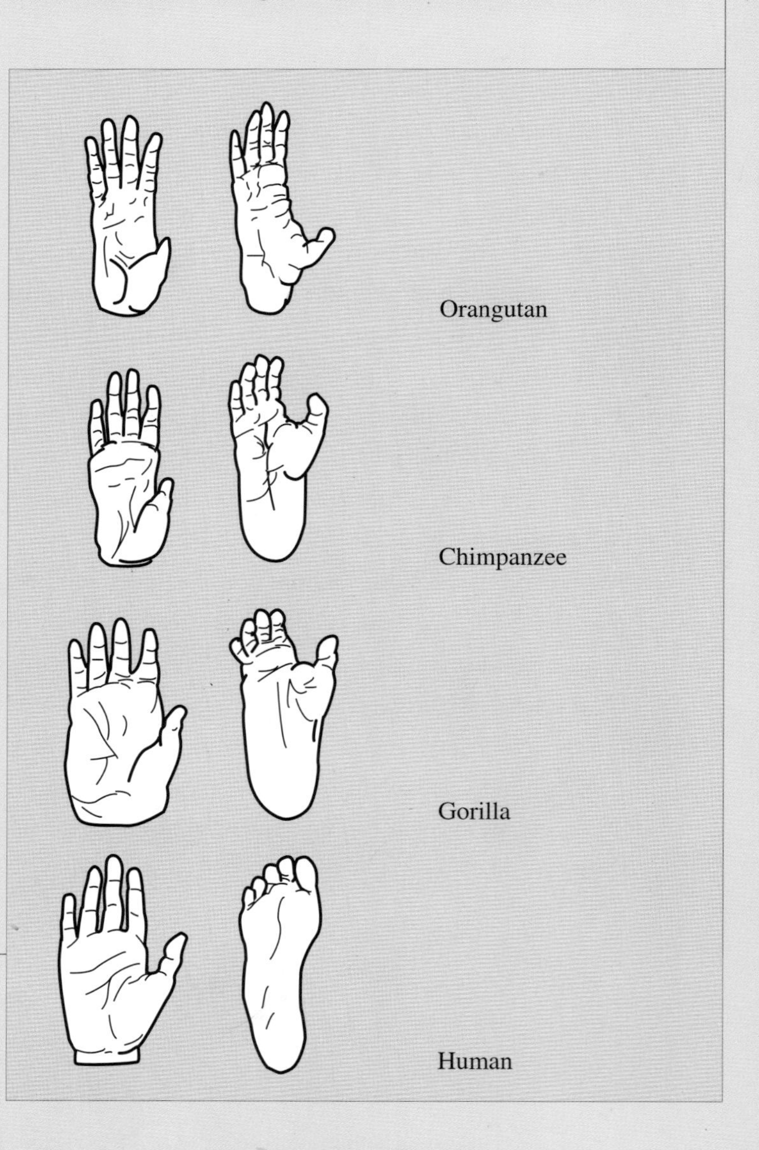

Darwin's other declaration—that our minds are equally similar—is more difficult to assess. Few can gaze into the eyes of a silverback gorilla without suspecting he is looking back with shared curiosity. But how can we begin to understand what goes on behind his wondering eyes? We can measure the apes' intelligence with human-devised tests, observe their reactions to human-induced stimuli, and even teach them to communicate with humans using computers or hand gestures, but these tests are based solely on our own "species-centric" ideas about what a mind can and should do. No test can fully reveal the intelligence and emotion of a great ape, but what these studies have uncovered are the enormous mental capabilities of chimpanzees, bonobos, gorillas, and orangutans—prompting us to continue examining our self-imposed separation from the animal world.

below: *Resting in a temporary day nest, a chimpanzee strikes the relaxed pose of contentment. Early research on great ape intelligence pulled the animals out of the wild and into laboratory settings to undergo human-devised tests. Yet Jane Goodall and those who followed her discovered more about apes' capabilities and their complex societies by observing them in the wild.*

THE APES DISCOVERED

The mention of manlike beings—"some animals [that share] the properties of man"—dates back to Aristotle, who lived between 384 and 322 BC. Aristotle had more than likely based his opinion on seeing the macaque, not the great apes.

European firsthand accounts of great apes did not occur until the late seventeenth and early eighteenth centuries, when explorers and naturalists returned home from Africa with tales of vicious manlike beasts. In the beginning of the seventeenth century, an English sailor described two strange beasts in Angola: "in all proportion like a man, but that he is more like a giant in stature than a man; for he is very tall and hath a man's face, hollow eyed, with long haire upon his brows... They goe many together and kill many negroes that travaile in the woods."

Much confusion surrounded these early discoveries and descriptions. In 1821 a noted authority described chimpanzee anatomy in detail, calling the animal an "orang-outan," a Malay word meaning "man of the forest." In the mid-nineteenth century, a missionary and an anatomist reported on "Troglodytes Gorilla, a New Species

above: *Two curious Sumatran orangutans enjoy a rare moment of companionship. Because of their solitary and arboreal habits, orangutans were the last of the great apes to be discovered by Western scientists and are the most difficult to track.*

opposite: *A young mountain gorilla looks out from the undergrowth with apprehension, a far cry from the vicious beast that early European explorers reported.*

above: *The expressive face of the adult chimpanzee is easily identified today. Yet early explorers struggled with distinguishing the many "manlike beasts" they encountered. Chimpanzees were labeled as young gorillas or orangutans, and some gorilla specimens were thought to be a subspecies of chimp.*

of Orang from the Gaboon River" in the *Boston Journal of Natural History*. Early specimens of chimpanzee brought back to Europe were believed to be a variety of gorilla, and the first young gorilla to reach Europe alive was labeled as a chimpanzee.

The confusion among the great apes illustrates their most common feature: their appearance and, at times, behavior are hauntingly and undeniably humanlike, a fact well known and accepted by the native people who shared the great apes' territory long before the European Age of Discovery. Indeed, just as the Malaysian name for the Asian great ape means "man of the forest," the term "chimpanzee" is thought to derive from a native African word meaning "mock man," and "gorilla" is, some suggest, an African name meaning "wild and hairy man."

Attempts to truly understand these manlike beings began in earnest in the early twentieth century, when Wolfgang Köhler began his study of chimpanzee intelligence at a great ape research station in the Canary Islands and Robert and Ada Yerkes set up a primate laboratory at Yale University.

Other scientists at this time were venturing for two to three months into the wild to observe the great apes in their natural habitats, but it is only since the 1950s that we have begun to truly appreciate and understand the significance of the great apes, their complex social structures, their keen intelligence, and their fragile hold on an ever-diminishing habitat. In 1959, American George Schaller began the first scientific study of great apes in the wild when he ventured into Africa to understand the life of gorillas. Though his study lasted only two years, and many studies have gone on to uncover more about wild gorilla behavior, Schaller was the first to begin unraveling the mystery surrounding the great apes.

The impetus to study the great apes in depth came from famed paleontologist Louis Leakey, who in the late 1950s was searching for artifacts of early man. He believed that learning more about the culture and behavior of early man would make him better able to look for and understand signs of their existence. With the declaration that "behavior doesn't fossilize," Leakey sent three women into the field to study the little-known lives of the great apes, hoping that their work would shed light on the behavior of prehistoric humans. In 1960 Jane Goodall began her groundbreaking study of chimpanzees in Gombe Stream, Tanzania (which was at that time called Tanganyika). Dian Fossey first laid eyes on the mountain gorillas of Rwanda and Zaire in 1966. And Birute Galdikas trekked into the forests of Borneo to study the elusive orangutans in 1972. Little could Leakey have known that these women would redefine the very meaning of "man" and "ape," observing behaviors previously thought to be exclusively human.

Leakey's protégées were chosen because of their enthusiasm and persistence; their lack of scientific training, which he felt might taint their findings; and their gender—women, he believed, would be viewed as less of a threat by the patriarchal communities of great apes. Regardless of his reasoning, Leakey obviously chose well. All three studies continue to this day and each of these women has inspired other individuals to carry on and expand the research, leading to remarkable new discoveries and insights into the lives of our closest relatives.

below: *A young bonobo swings playfully from a palm leaf, one of many natural playthings he will discover during his childhood. Bonobos were originally labeled a subspecies of the common chimpanzee* (Pan troglodytes)*, but were identified as a separate species* (Pan paniscus) *in 1929.*

Who Are the Great Apes?

Ongoing exploration of great ape societies and behaviors not only uncovers the similarities among these species, but also highlights their differences—from diet and social structure to territoriality and hierarchy. Just as we humans develop different customs, cultures, and even survival mechanisms to adapt to our place on the globe, the great apes—though strikingly similar to each other and humans in genetic makeup—lead surprisingly diverse lives.

All four great ape species live mainly along the equatorial rain forest belt—the orangutan in Asia and the bonobo, chimpanzee, and gorilla in Africa. The three African species share the most in terms of habitat and lifestyle. Because of the common chimpanzee's adaptability to various habitats—humid forest, deciduous woodland, or mixed savanna—its range is much broader than the two other species of African apes. The common chimpanzee lives in Western and Central Africa, the gorilla in the tropical secondary forest of Central Africa (mainly Uganda and the Democratic Republic of Congo [formerly Zaire]), and the bonobo in the tiny stretch of forest along the Zaire River in Congo.

right: *A gorilla mother shelters herself and her infant through an afternoon rain shower. Lowland gorillas live in lowland swamp and secondary forests throughout the equatorial forest belt of Africa. They forage and rest during the day and spend each night in a newly built nest of leaves and branches in the trees or forest shrubbery.*

previous pages: *A mother chimpanzee coddles her newborn infant. One of Jane Goodall's earliest discoveries was that most chimpanzee mothers are attentive, nurturing, and protective of their offspring until they are at least five years old.*

The orangutan is the only great ape to live in Asia, ranging the islands of Borneo and Sumatra. It also holds the distinction as the only great ape that lives mainly in the trees, maneuvering its large, red body through the treetops with surprising agility. The African apes, although fully capable of tree travel, spend much of their day on the ground, using closed fists to "knuckle-walk" through the dense undergrowth of the forests. Of the African apes, the bonobo is the most adept at bipedal (two-leg) locomotion. And because of the bonobo's slender build and structural ability for upright posture, a standing bonobo bears an uncanny resemblance to artists' interpretations of early man.

above: *An orangutan makes its way from one branch to the next with stunning precision and grace. Orangutans are the only great ape that live almost exclusively in the trees, a characteristic that makes them difficult to follow and study.*

above: *Highly social creatures, chimpanzees use facial expressions, vocalizations, body language, and gestures such as kissing and patting to communicate with other members of their group.*

Chimpanzees and Bonobos: The Closest of Cousins

Thanks to primatologist Jane Goodall's pioneering research, it is the chimpanzee that first opened our minds and imaginations to the remarkable emotional and intellectual capabilities of the great apes. Chimpanzees are engineers, toolmakers, and savvy negotiators. They form hunting parties to track and corner prey; communicate their needs and wants with sophisticated, often humanlike hand gestures and facial expressions; and, perhaps most remarkably, they exhibit the capacity to feel and show emotions much like what we term as fear, sorrow, anger, and joy. Chimpanzees, more than any other species, have helped to blur the line between human and nonhuman primates.

jane goodall, b. 1934

Even as a young girl in England, Jane Goodall knew she wanted to go to Africa. She dreamed of living like Dr. Dolittle and Tarzan—the characters in her favorite books—among the animals. And from a very young age she found herself fascinated with animal behavior, gathering worms to study their movements and hiding in a chicken coop to watch a hen lay an egg. She read book after book about the animals in the world, and assured her supportive mother that someday her dream would come true.

When Jane was twenty-three years old a friend living in Kenya invited her for a visit. This, Jane believed, could be the opportunity she'd waited for. In 1957, after months of preparation and saving, Jane boarded a boat bound for Kenya. Her life would never be the same.

It was in Kenya that Jane met the famed anthropologist and paleontologist Dr. Louis Leakey. Leakey was instantly impressed with her enthusiasm and knowledge of animal behavior. He invited her to assist him on an archaeological dig at Olduvai Gorge, where he was searching for signs of early humans. And it was there, on the magical African plain Jane had dreamed of as a child, that Leakey asked her to begin the unprecedented study of humans' closest relative, the chimpanzee.

In 1960, with no college degree or understanding of scientific methods, Jane Goodall and her mother, Vanne, stepped off the boat from Lake Tanganyika onto Gombe Stream Reserve in what was then known as Tanganyika (now Tanzania). Jane rattled the scientific community with her insistence on giving the chimpanzees names, rather than numbers, and recognizing their capabilities for feeling and showing emotion, intelligence, and reason. But when she began to file reports on findings such as tool use, hunting, and sophisticated communication, her critics could no longer deny the significance of her study.

With the encouragement of Louis Leakey, Goodall left her beloved Gombe for short periods to pursue a doctorate in ethology (the study of animal behavior) at Cambridge University in England. She received her Ph.D. in 1965.

Nearly forty years later, the study that many people predicted would last only a few months continues. The Gombe Stream Research Centre is staffed by local Tanzanians trained in studying chimpanzee behavior and tracking, and several permanent and visiting primatologists continue Goodall's study. Only one individual who was part of the original group of chimpanzees still survives: Fifi, the daughter of strong and capable Flo.

Although Goodall still visits Gombe several times a year, much of her life is now spent traveling the globe to raise awareness of the plight of the chimpanzees and to promote the conservation and environmental educational work of the institute that bears her name. In addition to continuing the behavioral research of the chimpanzees of Gombe, the Jane Goodall Institute, established in 1977, promotes the behavioral study and welfare of chimpanzees in zoos and other captive settings, operates four sanctuaries for orphaned chimpanzees throughout Central Africa, runs a reforestation and education program in Tanzania, and promotes care and concern for the environment, animals, and the human community through Roots & Shoots, a comprehensive environmental and humanitarian education program for young people around the world.

right: *Although bonobos are often confused with chimpanzees, they do have distinguishing characteristics. Bonobos have smaller heads, their faces are generally darker, they have tufts of hair that stand out from the sides of their faces, and they show an obvious hair part down the middle of their heads.*

Bonobos and chimpanzees are the most closely related species among the great apes, and the two that continue to cause the most confusion. Indeed, the two are so similar in appearance that bonobos were only identified as a separate species in 1929; before that date they were merely labeled as a subspecies of chimpanzee. Yet to the trained eye, the differences are easily identified. Bonobos have a slightly smaller head than the common chimpanzees; they have an obvious hair part down the middle of their skull; they tend to have reddish lips; and their faces are blacker than their lighter-skinned cousins. Although the often-used label of "pygmy chimpanzee" is inaccurate—for they do not differ much from common chimpanzees in height or weight—bonobos do have a more graceful and slender build.

Common chimpanzees and bonobos also differ in their group makeup and culture. While both species form communities of males and females that divide into smaller parties throughout the day, bonobo alliances are stronger and more stable. Also, unlike the alpha male–led chimpanzee communities, bonobo societies center around females, who make up the stable core of each group. Bonobos are an intensely peace-loving and gregarious species, avoiding social discord with a warm touch, embrace, or even sexual intercourse. Both species use grooming, touch, and social bonds to maintain peace and resolve conflicts.

The bonobo's delayed entrance into the primatological field of study is due in part to this creature's restricted and remote habitat. Sustained political unrest throughout Congo (the former Zaire)—a country that has at least two hundred ethnic groups, many of which have continued to struggle violently for control since the country's independence from Belgian rule—also deterred researchers from venturing into the forests to study the elusive, dark-faced "chimpanzee." In 1974, fourteen years after Goodall began studying the chimpanzees of Gombe, two field studies of bonobos were launched—one by a young couple named Noel and Alison Badrian and another by Japanese researcher Takayoshi Kano. It did not take long before the two teams of researchers discovered that this strange-looking ape had a social organization and culture all its own—from the female-dominated communities to sexually centered peacemaking.

Gorillas: Gentle Giants

Despite the early confusion, gorillas are easily distinguishable from chimpanzees and bonobos. For one thing, the gorilla is much larger—it has longer arms, broader hands, a larger skull, and a more pronounced brow ridge. In fact, the gorilla is the largest of the living primates. The species *Gorilla gorilla* is divided into three subspecies—the western lowland, eastern lowland, and mountain gorilla. All three subspecies of gorilla are similar in size. The western lowland sports a brown-gray coat; the eastern lowland a black coat, a longer face, and a broader chest; and the

below: *A silverback mountain gorilla takes a midafternoon rest. Gorilla males are the largest of the great apes, weighing up to 400 pounds (182kg). It is easy to understand how early encounters with these massive creatures could have turned to tales of monsters and man-killers, but research has proved that gorillas are simple and peaceful fruit eaters.*

dian fossey, 1932–1985

On a remote hillside in Rwanda's Parc National des Volcans stands a collection of wooden crosses marking the graves of mountain gorillas who lost their lives to poachers. Nestled among the crosses carved with the names Digit, Uncle Bert, Kweli, and Frito is the cross that pays tribute to the woman who died to save them. "Nyirmachabelli," *the sign reads—"the woman who lives alone on the mountain."*

Dian Fossey had a strong interest in animals from an early age, and began her college studies as a pre-veterinary student. But she switched her attention to occupational therapy midway through and worked in this field for ten years in Louisville, Kentucky. In 1963, perhaps to rekindle her early passion, she took a six-week vacation to Africa, where she joined other tourists on a gorilla safari and met paleontologist and anthropologist Dr. Louis Leakey at his research site at Olduvai Gorge. These two events would change her life forever.

When Leakey stopped in Louisville during a speaking tour in 1966, he remembered the tall, striking woman he'd met three years before in Africa. The next morning, the two had breakfast to discuss her dream of returning to Africa and his search for someone like Jane Goodall to study the mountain gorilla. Although Leakey did not offer Dian the job that day, he suggested that before she go to Africa she have her appendix removed, a precaution he recommended because his wife, Mary, nearly died after suffering an attack of appendicitis while far from medical help. Many wonder whether the suggestion was merely a test of Dian's true commitment. Nevertheless, three months after their first encounter, Dian had a preemptive appendectomy in preparation for life in the bush. Four months later she left for Africa.

In 1967 Fossey established Karisoke research camp in the Parc National des Volcans. In the early months she followed the research methods of George Schaller, a respected zoologist and anthropologist who had spent two years observing gorillas in the wild from cautious distances, careful not to be seen by the shy, gentle giants. But soon Fossey began to inch her way into the gorillas' comfort zones, allowing them to become aware of and acclimated to her presence.

This method made history in 1970 when Peanuts, an adult male gorilla, touched Dian's hand. With the trust of the gorilla group, Dian logged thousands of hours studying their feeding behaviors, their mother-infant bonds, and their peaceful, quiet lifestyle. In 1978, Digit—the silverback gorilla she had known since he was a "playful little ball of disorganized black fluff" and the star of the National Geographic *television special—was murdered and beheaded by poachers. From that day on Dian waged a bitter war against those who would dare to kill the mountain gorillas.*

She obtained her Ph.D. from Cambridge University in 1980 and went on to write about her work in the book Gorillas in the Mist, *which later became a movie starring Sigourney Weaver. With the outpouring of support after Digit's death, Dian established the Digit Fund to help protect the remaining gorillas.*

In 1985 Dian was murdered in her cabin at Karisoke. The crime remains unsolved, yet her memory and dedication to the protection of the mountain gorillas lives on in the Dian Fossey Gorilla Fund (formerly the Digit Fund), which promotes the ongoing study of mountain gorilla behavior and the conservation of the species and its habitat.

left: *A young western lowland gorilla takes comfort in his mother's arms. Gorillas travel in troops made up of one male silverback and several females and their young. When adolescent gorilla males become sexually mature, they leave their natal group to start their own troop.*

mountain gorilla has longer black hair, with a larger jaw and shorter arms. The two subspecies of lowland gorilla are separated geographically by the Zaire River.

Early explorers' reports of the savage gorilla bestowed upon this great ape an unfair reputation as a ferocious man-killer. Indeed, many who first encountered the gorilla in the wild were in search of trophies rather than data. It is no wonder they came back with tales of chest-beating giants with massive canine teeth and arms that could tear a man limb from limb. The gorillas these adventurers encountered—most likely the alpha male silverbacks—were on the defense, trying in vain to protect themselves and their troops from the threat of intruders.

When researcher Dian Fossey and others began to venture into the gorillas' world with binoculars and note pads rather than nets and shotguns, we came to see an entirely different animal. Fossey revealed societies of gentle and peaceful beings who ask for little more in life than the right to forage quietly through brush and rest undisturbed under the lush rain forest canopy.

Gorillas live in troops of approximately twelve individuals, made up of one silverback male and a group of females and their young. The gorilla has the most stable grouping pattern of the great apes, with the same individuals traveling together for as long as several years at a time. Group ranges overlap significantly, but there is little territorial defense among the communities and aggression within a group is rare. In fact, male gorillas become aggressive only when trying to attract females from other groups. As the largest primate alive, the gorilla truly is the gentle giant of the African forest.

Orangutans: Elusive Tree-Dwellers

The orangutan is a solitary creature, both in its lifestyle and in its relation to the other great apes. It is the only great ape that resides in Asia and the only one that spends the majority of its day high in the trees. The orangutan received less attention in the early years of great ape study because, as a tree-dweller in Asia, it was thought to have less significance in the search for human origins than its ground-dwelling African cousins.

Its reputation and significance among the locals of Borneo and Sumatra, however, go back many years. Orangutan skulls have been found in prehistoric caves, pointing to the ape's target as prey by prehistoric man or to its popularity in religious rituals, still practiced among some present-day peoples. The Dyak of Borneo, for example, revere orangutans as spirit heroes and treat their skulls as sacred objects. By contrast, during the eighteenth and nineteenth centuries the orangutan gained a less reverential reputation—that of kidnapper and molester of young human women. And many horrid tales of the orangutan's lust for humans followed explorers from the villages of Indonesia to the shores of Europe and into the early annals of natural history.

In truth, the orangutan is neither saint nor seducer. It is a striking creature with brilliant red hair that hangs recklessly around its bare face and flows from beneath

right: *The orangutan, the only great ape to reside in Asia, is also distinguished from other great apes by its coat of red hair. Sumatran orangutans like this one have a paler and longer coat than their cousins in Borneo.*

its arms like fringe on a jacket. The adult male orangutan is significantly larger than the female, with large throat pouches that look like double chins and huge cheek flaps that extend from the sides of the face. Bornean males have larger flaps on their faces than those of Sumatra. And Sumatran orangutans are more slender than the Bornean subspecies.

Adult orangutans travel alone, spending their days high in the trees with only occasional interaction or association with others. Their solitary lifestyle and limited vocalizations—other than the adult male's "long call" to alert others of his presence—made early field studies painstaking and frustrating. Once Birute Galdikas and others were able to find, identify, and follow the mysterious red ape, they began to realize that the orang's elusive nature is necessary for its survival, and that what first appeared to be random, isolated meandering through the trees is a well-mapped journey from one food source to the next. Galdikas and her colleagues were also the first to realize that, due to large-scale commercial logging and widespread poaching, the orangutan claimed the unenviable title of the most highly endangered species of great ape—a distinction that narrows each year as the African great ape populations continue to decline.

birute galdikas, b. 1946

Jane Goodall and Dian Fossey were already famous for their groundbreaking studies of chimpanzees and gorillas when Birute Galdikas became the third of Louis Leakey's protégées. Finishing her master's degree in anthropology at the University of California at Los Angeles, she had already decided that, upon graduation, she would earn money as an archaeologist and use it to fund an independent study of the wild orangutans of Indonesia.

When Galdikas attended a lecture by Leakey in 1969 she became convinced that he was the man who could help her fulfill her dream. She approached him after the lecture to explain her desire and met with him again the following morning for an interview. Because Leakey's planned field studies of the chimpanzee and gorilla were well under way, he was ready to begin the study of the third great ape, continuing his search for insights into early humans. Little did he realize that, like Goodall and Fossey before her, Galdikas had been preparing for just such a position all her life.

Birute Galdikas was born in Wiesbaden, Germany, to Lithuanian parents. When she was two years old, she moved with her parents, sister, and two brothers to Ontario, Canada, where her father worked as a miner and her mother as a nurse. It was in these early years that she learned the valuable skill of crossing barriers into new and unfamiliar cultures—a talent that would serve her well as she entered the lives of the Indonesian people and the shy red apes of the forest. Birute grew up fascinated by science, history, and nature, and could listen for hours as her mother told stories of the history of mankind and the progress of civilization.

Finally, in 1972, Galdikas trekked into Tanjung Puting Forest in Borneo, carrying with her only the contents of two large backpacks. There she set up Camp Leakey, named in honor of the man who sent her there, to study the elusive orangutans.

Galdikas submitted her 333-page Ph.D. dissertation to UCLA in 1976, documenting five years of intensive field study of orangutans, noting their eight-year birth interval, their solitary lifestyle, and the vast variety of fruits and vegetables in their diet.

As Galdikas's research continued, she became aware of the growing number of infant orangutans taken from the forests for pets. The fact that the hunters first must kill the mothers, and often other orangutans as well, in their search for marketable infants the pet trade posed a severe threat to the long-term survival of the species. Galdikas convinced the Indonesian government to confiscate the infants and turn them over to her for care and rehabilitation. Soon Camp Leakey was overrun with young orangutans who looked upon Galdikas as their caregiver and protector. Few can forget the National Geographic *photograph featuring a young Birute with an infant orangutan in her arms and two others clinging to each of her legs.*

Since then she has divided her attention between continuing research of the wild orangutans and the rescue and rehabilitation of orangutans in peril. In 1986 she established Orangutan Foundation International, based in Los Angeles, California, to support her objectives. OFI also focuses on the restoration of the degraded forest around Tanjung Puting National Park and education on the plight of wild and captive orangutans around the world.

Their Future in Our Hands

The unique characteristics of each species of great ape serve to emphasize these creatures' ability to adapt to the needs and challenges of their particular environments. Yet the ever-expanding human population and our continuing exploitation of the earth and its nonhuman inhabitants is one challenge they may not be able to surmount.

Throughout history, people have judged the great apes not on their own merits but as human precursors or as somewhat deficient stand-ins for humans. They have been injected with human diseases, shot into space, and stolen from their mothers to serve as pets or surrogate children. They are paraded under the big top of circus tents and trained to perform for cameras in television commercials and big-screen movies. Their forest homes are logged by commercial timber companies or burned down to make room for subsistence farmland.

below: *Virunga Volcanoes Park in Rwanda is one of the few protected habitats remaining for great apes, either in Africa or Asia. Much of their forest homeland is under attack by the ever-growing human population and rapidly expanding commercial logging operations that bulldoze dirt roads into territory once untouched by human hands.*

When we take the time to look at the lives of great apes in the wild, we see that Darwin's proposal that we are descendants of apes is less earth-shattering than it is irrelevant. When we replace fear and repugnance with curiosity and respect, we begin to see great apes not as human surrogates or runners-up in the evolutionary showdown, but as complex, sentient beings who reason and solve problems, feel and show emotions, and whose child-rearing and social skills could sometimes be used to instruct our own species.

With this knowledge, our emphasis must turn from observer to rescuer in order to save the great apes and their habitat from certain extinction. We must act not only because we have so much more to learn about their behavior, culture, and capabilities in the wild, but because allowing these noble and intelligent beings to disappear would be one of our own species' greatest shames.

right: *A morning mist blankets the Malaysian rain forest, home to the Borneo species of orangutans. As in Sumatra, the precious habitat is under siege by timber companies, urban growth, and widespread forest fires.*

Ham, 1957(?)–1983

chimpanzee

In the late 1950s and early 1960s, hundreds of infant chimpanzees were recruited into the United States space program. Torn from their mothers' arms in the wilds of Africa, the chimpanzees were packed into crates and flown to Holloman Air Force Base in New Mexico, where they underwent a barrage of physical examinations and treatments for jungle diseases and parasites. All had to be under 50 pounds (23kg).

Once the unhealthy and oversized recruits were eliminated from the lineup, the tour of duty intensified. The chimpanzees underwent rocket sled rides, weightlessness and isolation experiments, reflex exams, and intelligence tests. In the end, only twenty animals were chosen. Their mission: to test the safety of space travel and pave the way for American astronauts into the frontiers of space. One of these "astrochimps" was a young male chimpanzee named Ham, short for Holloman Aeromedical.

To prepare the bewildered primates for their mission, the researchers at the Aeromedical Research Laboratory at Holloman devised a multitude of simulations and experiments. In one such test, the infant was placed in a tiny cubicle and strapped into a nylon mesh suit. In front of him was a panel of gadgets and knobs. After 1,093 hours of training, the chimpanzee was expected to react correctly. When a red light flashed, he was expected to press one lever; when the blue light flashed, he pressed another. If the chimp performed correctly he received a small pellet of food or a drop of water from a tube near his head. When he failed, an electric shock shot up through the pad on which he rested his feet. The chimpanzee was tested for several minutes, then rested several minutes. The entire test on the so-called "learning machine" lasted sixty-six minutes. The test was conducted eight times during one twenty-four-hour period.

When Ham completed his training, he was ready for his mission. On January 31, 1961, the three-year-old chimpanzee was strapped inside a tiny, egg-shaped capsule and blasted into the air on a Redstone rocket at 5,000 miles (8,000km) an hour, becoming the first U.S.-trained primate to enter space. The rocket was scheduled to fly 115 miles (184km) up into the air and 290 miles (464km) downrange at a speed of 4,200 mph (6,720kph). But the rocket malfunctioned. Ham sped 156 miles (250km) up and 414 miles (662km) downrange, reaching a top speed of 5,800 mph (9,280kph). Trained for a maximum speed of 11 Gs, at one point he pulled 18.

When Ham emerged from the capsule, after plummeting to the earth at unprecedented speeds and bobbing in the Atlantic Ocean for three hours because of the error in range, his face revealed the terror he was unable to express. His eyes were wide and searching, and his lips stretched back from his mouth baring his tightly clenched teeth. Ignorant reporters, believing he wore a smile, labeled him the "grinning hero." Chimpanzee expert Jane Goodall labels it the most intense fear grimace she has ever seen.

After the historic flight, Ham adorned the cover of Life *magazine, but soon the attention of doctors, reporters, and Air Force staff faded. Ham was no longer needed. For sixteen years he lived alone at Washington, D.C.'s National Zoo. He at last joined other chimpanzees at the North Carolina Zoological Park in 1980. Ham died of an enlarged heart in 1983 at the tragically young age of twenty-six.*

chapter ii

growing up in the forests

Great apes begin their lives much as humans do—tucked safely in their mother's arms. Each day the great ape mother shelters her infant from danger, teaches him to communicate, and instills in him an understanding of right and wrong. His childhood is long, for he has much to learn. And if he loses his mother before he has reached adulthood, his life is forever changed.

No other nonhuman primate has a childhood as prolonged or that requires as much learning as the great apes'. Their mother-infant bonds and long-term family relationships are strikingly similar to our own. Although each great ape species has a unique learning curve and complex code of conduct that guides them to adulthood, the one constant in each great ape's childhood is none other than dear old mom.

Like humans, great apes generally have only one child at a time. When the rare multiple births (usually of twins) do occur, the infants are so small and the mother so weak that often one of the infants does not survive. Because single-child births are the norm, mothers can provide one-on-one attention as they teach, nurture, and shelter each child until he reaches full maturity.

All great ape newborns possess the telltale signs of infancy: round faces, pudgy noses, and awkward, uncoordinated movements. Perhaps for this reason, just as in humans, infants customarily receive a great deal of attention from other group members or older siblings. When a chimpanzee is born, female members of the group gather eagerly around the newborn to catch a glimpse. And older siblings try to touch, hold, or grab the new family member—always under the watchful eye of the cautious mother.

Enveloped in his mother's arms, a new gorilla, weighing only four to five pounds (1.8–2.3kg), peers out at the world around him with wondering black eyes framed in a grayish pink face. For now, he is sparsely covered with fur, but soon he will develop a full coat of hair and a small white "tail" tuft that he'll retain throughout childhood, signifying to others that he is still a youngster.

White tufts of hair also adorn the rumps of young chimpanzees and bonobos, again as a sign of their age and to help their mothers identify them in a group of other fluffy-haired primates. Chimpanzee infants, weighing approximately four

opposite: *A mother Sumatran orangutan ambles through the forest with her fifteen-week-old offspring napping on her back. Young orangutans face their first few weeks clutching onto their mother's belly, venturing to the less-secure back position only when they have mastered the art of balance.*

pounds (1.8 kg) at birth, are born with light pink faces and ears that darken with age. Bonobo newborns, who weigh around three pounds (1.4kg), begin life with the same darkened face as the adults.

The orangutan infant, weighing two to four pounds (0.9–1.8kg) at birth, differs from adults in the bright red hair that stands on end on the crown of his head and the pink coloration around his eyes and mouth.

Orangutans give birth approximately every seven or eight years, giving the orang youngster the longest childhood of the four great apes. Gorillas and chimpanzees give birth approximately every five or six years, and bonobos about every four years. But even after another child is born the elder child in all four species stays near mother's side until he is able to face the world on his own—an experience that requires years of maternal preparation.

right: *Mother bonobo gently kisses her infant, a poignant illustration of the maternal devotion exhibited by the great apes for their young.*

great ape facts

gorilla

GESTATION: Approximately 250 days
AGE OF SEXUAL MATURITY: Male 11–13; Female 7–8
ADULT WEIGHT: Male up to 400 lbs. (182kg); Female up to 200 lbs. (91kg)
LONGEVITY IN THE WILD: 40 years

chimpanzee

GESTATION: Approximately 230 days
AGE OF SEXUAL MATURITY: Male 10; Female 8
ADULT WEIGHT: Male up to 110 lbs. (50kg); Female up to 85 lbs. (39kg)
LONGEVITY IN THE WILD: 50 years

orangutan

GESTATION: Approximately 250 days
AGE OF SEXUAL MATURITY: Male 15; Female 12
ADULT WEIGHT: Male up to 200 lbs. (91kg); Female up to 110 lbs. (50kg)
LONGEVITY IN THE WILD: 50 to 60 years

bonobo

GESTATION: Approximately 230 days
AGE OF SEXUAL MATURITY: Male 10; Female 8
ADULT WEIGHT: Male up to 100 lbs. (45kg); Female up to 75 lbs. (34kg)
LONGEVITY IN THE WILD: 40 years

right: *A chimpanzee infant clutches the hair on his mother's belly as she travels the savanna. Young chimpanzees ride in this ventral position until approximately six months of age, when they move to their mothers' back.*

opposite: *Young chimps typically stay close to their mothers, and don't venture out into the trees until they are at least four years old.*

The Early Years

From the moment a great ape is born he is melded to his mother, cradled in the safety of her lap or pressed against her belly. For the first few weeks she will support him there with one hand or with the pressure of her thighs as she moves from food source to resting place. Soon the baby develops the strength to cling to his mother by clutching fistsfuls of hair with all his fingers and toes. His grip must be tight and firm, because his mother is likely to move at any time—to transfer from one branch to the next in search of food or run out of the way of an angry male.

As the young apes grow older they develop the coordination that allows them to move to their mothers' back for transportation. Orangutan babies ride on their mothers' back until they have mastered the art of brachiation (swinging from one branch to another), at about the age of three. Their initial forays into the branches are clumsy and frightening. When mother and child approach a leap greater than the little one is able to make, the mother orangutan often holds the tree limbs together, narrowing the gap, or she may use her own body as a bridge for her young one to cross.

Gorillas also depend on their mother for transportation until they are about three years old, by which time they are fully able to travel on their own. A gorilla's first unsteady steps from mother's back begin at around five or six months. When he is eighteen months old he'll begin to follow his mother on foot, always staying close by in case he gets frightened, startled, or too weary to carry on.

Chimpanzees and bonobos move from the ventral position (clutching mother's stomach) to riding like a jockey on her back at about six months old, and stay there until at least the age of four.

Flo, 1929 (?)–1972

chimpanzee

One of the first chimpanzees to accept Jane Goodall's presence in the wild forest of Gombe was a grand old female Goodall named Flo. When Jane first laid eyes on Flo in 1962 she saw a scruffy, tired chimpanzee with tattered ears, worn teeth, and a bulbous nose. But as time went on Jane saw a different chimpanzee. Unlike other older females, Flo was relaxed with adult males, stood up for herself in community scuffles, and always won out among other females when tests of will ensued for food. She was, as Goodall soon realized, the dominant female of Gombe's Kasakela community. This high status was not won through warfare and intimidation, as with males. Flo became a high-ranking female because of her exceptional personality, her social polish, and her assertive and self-assured manners—valuable traits she passed on to her children, and which now live on in her grandchildren.

In 1962 Flo was the only adult female to brave the unfamiliar territory of Jane's camp in search of bananas. She came carrying an infant female, whom Jane named Fifi, and accompanied by her juvenile son, named Figan. And Jane soon surmised that there was yet an older son, Faban. Watching Flo interact with her family was magical. Flo was attentive and playful, disciplining the children when necessary but never aggressively. When Flint was born in 1964 Flo, though even older and growing weary, provided him with as much attention and nurturing as she had Fifi and, presumably, Figan and Faban. But when it came time to wean young Flint, she did not have the energy to resist his demands. Flint refused to move through the passage of maturity. Even when she gave birth to Flame, Flint joined them in their nest at night and sometimes clutched onto her belly—side by side with Flame—to travel with her through the forest. Flo, probably over forty years old, simply could not handle the double burden. And when she fell ill, unable to find the strength to climb into a tree, Flame disappeared. Perhaps because her load was lightened, Flo's condition improved. But Flint's abnormal insistence on ongoing attention and care persisted. At times Jane believed Flo depended as much on Flint's company as he on hers. Fifi, now a curious juvenile, did not often travel with her mother, and Flo was too slow to keep up with many of the other members of the community. One day, when Flo and Flint approached a fork in the path, they took separate trails. Flo looked to see if Flint had followed her and, realizing that he hadn't, turned around to join him.

One morning, eleven years after they had met, Jane was told that Flo had died, her body found lying face down in a stream. "Even if I had not arrived to record her history," Jane writes in Through a Window, *"to invade the privacy of that rugged terrain, Flo's life would have been, in and of itself, significant and worthwhile, filled with purpose, vigour and love of life." Flo's obituary appeared in Britain's* Sunday Times, *yet it is her living legacy that will carry her memory through time. Her son Figan reigned as alpha male until 1975 and Fifi now holds her mother's position as the dominant female of the Kasakela community. Fifi has seven healthy, high-ranking children—Freud, former alpha male, Frodo, current alpha male, Fanni, Flossie, Ferdinand, Faustino, and Flirt. Grandma Flo would be proud.*

left: *Orangutans, who stay with their mothers until they are at least seven, have the longest childhood of any of the great apes. This is likely because, as semisolitary animals, they must learn all the lessons of survival before setting off into the forests for a life on their own.*

By watching their mothers' technique and through many episodes of trial and error, the young apes are soon swinging and walking alongside their mothers. Yet great ape mothers are not merely modes of transportation for their curious offspring. Great ape mothers discipline, nurture, and teach their infants, strengthening the bond and helping to prepare them for a life on their own.

One of the most remarkable characters Goodall studied in her early years in Tanzania was an aging female she named Flo, the matriarch of the chimpanzee community. Flo was the mother of five known chimpanzees, all of whom maintained a high status in the community, thanks no doubt to her superior mothering skills. What was it about Flo that produced such well-adjusted and high-ranking offspring? Genetics surely play a part—good health, an intelligent mind, and keen common sense. But one cannot deny the importance of a good upbringing.

"Old Flo was a highly competent mother, affectionate, tolerant, playful and protective," wrote Goodall in *Through a Window*. "She spent a good deal of time with other members of the community, and she had a relaxed and friendly rela-

below: *Sticks, rocks, vegetation, and other objects in the environment are the educational toys of great ape youth; by playing with these things the young ape develops necessary motor skills and learns the nature of his surroundings.*

tionship with most of the adult males. In this social environment, Fifi [Flo's only daughter] became a self-confident and assertive child."

Flo established close bonds with each of her infants and helped them develop their motor skills and social graces through tender sessions of grooming and spirited bouts of play. When her infant son Flint was but a few months old, she would dangle him gently above her head with one hand and softly tickle him with the other as his lower lip hung loose in a chimpanzee smile. And when Flint was older and would chase his sister around as Flo lay resting in the grass, she'd occasionally grab a passing foot to join in the fun.

Play is an important part of many animals' childhood, and because of the complex survival skills great apes must learn, it is especially critical for them. The early interaction such as tickling and play biting between mother and child encourages motor development and coordination. Great ape infants also entertain themselves with their own form of play and exploration. Just as with human babies, great ape infants become fascinated with their own fingers and toes, as well as various mouth-sized inanimate objects within their reach—leaves, sticks, or the food their mother has in her mouth.

above: *Orangutan youngsters imitate their mothers' nest-building and eating habits during play, building on the skills they'll need later as adults.*

Young chimps roll stones or small fruits on the ground and toss them high into the air, then retrieve them. They use sticks or stones to tickle themselves in hard-to-reach places. As they grow older, they invent elaborate games of tree climbing, swinging, twirling, and semibrachiating in trees. Although it's all fun and games to them—signified by the telltale open-mouth play face and breathy laughter—they are developing strength and valuable locomotion skills for later life.

Orangutan play involves swinging in the trees and their own version of playing house by constructing awkward night nests or toying with the food on which their mothers dine.

As the great apes grow older and are able to venture away from mother's reach they begin to experiment with social interaction. Although young orangutans must generally rely on their mother or an older sibling for playmates, they do have occasion to play with others their age when orang mothers gather at a fruit-bearing tree—a bit like the "play dates" of our own species.

Because they live in groups, gorilla, chimpanzee, and bonobo youngsters have a variety of ready playmates. Play among these apes includes wrestling and mock fighting, as well as games similar to follow the leader (sometimes interpreted as chase the leader!). Little must they realize that this rough play is actually a test of each other's physical strength and tolerance, information that will prove useful when they go on, as adults, to establish rank and status. It is not uncommon to see a group of young gorillas clambering over an even-tempered silverback, illustrating the true patience of the gorilla elder toward the white-tufted youth.

below: *Through play, socializing, and quiet grooming sessions with his mother, the young bonobo forms valuable allies and learns the social graces of bonobo society.*

opposite: *As chimpanzees grow up, they begin to spend more time in rambunctious games with their playmates. It is during these play sessions that a chimp starts to carve out his or her place in the complex social order.*

right: *When young gorillas are not engaged in social play with peers or adults, they invent their own games of swinging, climbing, and exploring—building strength and skills that will serve them well into adulthood.*

previous pages: *A silverback mountain gorilla dangles a youngster playfully above his head. Silverbacks are not only tolerant of the children in their troop, but often seek them out for play and grooming. Silverbacks have even been known to adopt infants whose mothers have died, allowing the orphans to sleep in their nests at night.*

But adults are not always understanding of the carefree exuberance and untamed curiosity of childhood. Much as a human mother must usher her child into social settings with care and discretion, a great ape mother—gorillas, chimpanzees, and bonobos in particular—must monitor her child's first interactions with other individuals and stand ready to step in if the child appears in danger of annoying an adult or causing a scene.

When a young gorilla totters curiously toward a silverback who is clearly agitated, for example, his mother intervenes by uttering soft grunts of warning or pulling him back to her side. A chimpanzee mother will scoop her playing child out of the way of a displaying male. And although an orangutan mother has fewer encounters with other orangs, her demeanor when other orangs are in the vicinity illustrates to the youngster that social interaction is generally not the norm.

Through trial and error, the great ape child learns which behaviors are appropriate around which individuals. He will also learn when and why it is necessary to act submissive, when to flee, and when merely staying close to mother is the safest bet of all.

Mother Knows Best

Just as social competence and appropriate responses to various situations are learned behaviors in both great apes and humans, so are many of the basic survival skills. A human child does not know instinctively that he should not eat the bright red berries on the Christmas holly, nor does a young orangutan know which plants in the forest are edible and when they are in season.

The learning process is gradual. The orangutan infant may not appear to be taking notes as he travels passively from tree to tree on his mother's back, but over the years he begins to understand which fruits are in season and where they are located. These data are especially important for an orangutan, who will eventually be traveling on his own, unlike the African great apes, who can continue to follow the group from one food source to another. But knowing where to find the food is sometimes only half the battle. Orangutans must also learn how to discard the prickly skin and spit out the seeds of the durian fruit and to use their teeth and hands to open the tough-skinned mangosteens.

below: *An infant orangutan kisses his mother during a moment of play and affection. Although orangutans require just as much care and nurturing during their formative years as other great apes, as adults they seem to prefer their solitude, interacting with others only on rare occasions.*

right: *Lowland gorilla youngsters have moved to a new stage in their childhood as they begin to sample the foods that will make up the majority of their lifelong diet. They learn from their mothers which foods are edible and when fruits are ripe.*

Many plants in the bonobo diet also have both edible and inedible parts. The young bonobo learns how and what to eat by sharing his mother's food and watching as portions are discarded. Gorilla infants learn as they observe their elders strip leaves from bamboo stalks or strip the hollow stem of wild celery before dining.

Of course, as toolmakers, chimpanzee infants have the most to learn about food preparation and retrieval. Making and using tools are skills they practice early and often, though initially as simple play. A young chimpanzee watching his mother fish for termites will pick up his own blade of grass and clumsily insert it into the mound with little success. But by doing so he begins to learn which blades or sticks are best and how they should be stripped, and he builds the coordination necessary for a smooth performance. Chimpanzees also break nuts using stones as a hammer and anvil, wad leaves to create crude sponges and napkins, and brandish large sticks to frighten or intimidate enemies—all skills requiring perception and practice.

After a long day of observation and play, the great ape infant sleeps nestled alongside his mother in her temporary nest, which she constructs each night in the trees—or lower bushes, for gorillas—by bending fresh branches and leaves into a springy, soft bed. As they grow older the youngsters begin to experiment with their own nest-making by carefully copying the method their mothers have used night after night. Yet few will venture away from the comfort and safety of mother's bed until a new infant is born, signifying a shift in the mother's focus and an end to the very special, one-on-one care.

The anticipation of a new sibling also marks the end of a young great ape's dependence on mother's milk. Gorillas suckle until they are approximately two years old. Chimpanzee mothers begin to wean their young at three, and orangutans at approximately six, again longer than any of the other great apes. After weaning, a young great ape will attempt to suckle when frightened or insecure, even when his mother is no longer able to produce milk. Once the new infant arrives, the mother's milk may be reserved for the newborn, but she is still sole caregiver and protector of the growing adolescent. Indeed, many a great ape mother travels with a newborn clutching her belly and an older infant riding securely on her back.

below: *Chimpanzee infants suckle until they are at least three years old, when their mothers begin to wean them—usually in preparation for another child. Yet the weaning can be difficult for some stubborn young chimpanzees, who often return to the security of mother's breast when frightened.*

right: *A young western lowland gorilla seems to contemplate life while she sits peacefully among the troop. Although she has several more years to go of tutelage from her mother, she will one day leave her natal group to join another troop or create a new one with a lone silverback.*

Leaving Home

Full independence from mother does not occur until the great ape reaches sexual maturity, when he can take with him the lessons from his youth and venture away from his mother's protection and guidance.

Gorilla females become sexually mature at about the age of eight, when they begin to show a slight genital swelling for two to five days each month. Soon thereafter they will leave their parental group to unite with solitary silverbacks or to join other existing troops, presumably to avoid inbreeding with their fathers. Male gorillas begin to mature at around eight—when they are referred to as blackbacks. They do not become sexually mature until they develop the regal silver coloring on their backs, between the ages of eleven and thirteen. Their full maturity is also marked by a pronounced sagittal head crest and a muscular body. At this time the male will also leave the group to travel alone, sometimes for years at a time, until he establishes his own troop. He will not begin to breed until he is between fifteen and twenty years old.

Orangutan females reach sexual maturity around the age of twelve, but will not bear their first offspring until fourteen or sixteen. Like the gorilla male's blackback phase, orangutan males experience what is known as a subadult phase between the ages of ten and fifteen. At fifteen, they become physically and socially mature, with full cheek pads, a large fatty nodule on the head, darker and longer hair, and a large throat pouch that enables them to make the deep, groan-like "long call" that warns nearby males of their presence. Although the reasons for the male's delayed maturity are not yet understood, researchers believe a young male matures more slowly when older, dominant males are in the vicinity.

left: *Orangutans spend the majority of their lives in the trees, moving their massive bodies from branch to branch with great agility. Their long arms, short legs, and hook-shaped hands and feet suggest a long history of life in the trees. This young Borneo orang has just begun to master the art of balance, grip, and grace.*

Young orangutan males leave their mothers' side immediately after the birth of a sibling to live a solitary life, but a young orang female stays with her mother a bit longer, playing with the new infant and learning the ropes of motherhood. Because adult females sometimes meet at fruiting trees, mother and daughter are likely to see each other again.

Female chimpanzees and bonobos become sexually mature at about seven or eight years of age, while males reach maturity a few years later, at about ten. As the females mature they show their first signs of a pink "bottom" swelling. This swelling—an indication of sexual receptivity—occurs every five weeks thereafter for chimpanzees; bonobo females can have the swelling more often and for longer periods. Although many zoogoers recoil in embarrassment or shock when first laying eyes on a female chimpanzee or bonobo in estrus, thinking the poor female has been injured or is deformed in some hideous way, the males of the species find the pink swellings provocative and enticing. Chimpanzee or bonobo females sporting a pink bottom become the belles of many a great ape ball.

right: *Running to mother for comfort and protection is common behavior for great ape young ones. And, just as with this vigilant western lowland gorilla, good mothers respond with concern and attention.*

These lovely pink swellings gradually get larger with each cycle over the years, until they attain their full size when the female chimpanzee reaches ten or eleven, at which time she is at her most attractive to the adult males in the group. Although chimpanzee males begin to show early signs of maturing at the age of seven or eight, they have several more years to grow until they reach their full adult weight and size, at approximately sixteen or seventeen years of age.

As with gorillas, chimpanzee and bonobo females leave their home, or natal, group to transfer to another community upon sexual maturity. On the rare occasions when the female chimpanzees do stay within the natal group—as with Flo's daughter Fifi and Fifi's daughter Fanni—the females maintain a lifelong relationship and it is not uncommon to see three generations traveling together from day to day.

Male chimpanzees and bonobos remain in the community in which they were born, developing close alliances with their brothers and often maintaining ongoing relationships with their mothers. Fraternal bonding is more prevalent among the chimpanzee communities, in which one brother may help another to establish his

alpha status. With bonobos, for whom the females form the core of the society, a mother's status in the group greatly affects her son's relationships with others. If a bonobo male is born to a low-ranking mother, his chances for achieving high-ranking status as an adult are slim.

The Life of an Orphan

The great apes' dependence on their mothers even after weaning—for ongoing learning, comfort, and caregiving—becomes sadly apparent when a youngster loses his mother before he reaches adulthood. Without the continuing tutelage and nurturing, many orphaned youngsters simply die, unable to fend for themselves in a world that demands long-term maternal care. When old Flo's life finally came to an end, her young son Flint, though physically able to survive on his own, could not cope with the loss of his constant companion. He stopped eating, grew listless, and curled up near a stream at the site of his mother's death, never to move again.

opposite: *Without the care and protection of his mother, this little orangutan would have scant hope of finding adequate food or successfully negotiating the swampy forests of his native Borneo.*

left: *This western lowland gorilla mother and infant stare intently with shared curiosity at insects on a twig. Gorillas, like all other great apes, learn their survival skills by watching and imitating their mothers and other community members.*

below: *This young orangutan seems to have few cares as he moves with ease through the steamy Borneo jungle, secure in the knowledge that he will be well cared for and watched over until he is mature enough to live on his own. But some orangs are not so lucky. Killing mother orangutans and stealing their infants as pets has become big business throughout Indonesia, further threatening the precarious future of the orangutan species and destroying the life of the young orangutan, who may never again see the forest he once called home.*

Countless examples—both in the wild and in laboratory settings—point to the behavioral disturbances experienced by nonhuman primates who have lost their mothers at an early age. Clinical depression, decreased frequency of play, and lethargy are just a few of the effects. Similar to the psychosocial dwarfism (slowed growth due to nonphysical factors) seen in human children who experience maternal deprivation, great ape orphans who survive, especially chimpanzees, often fail to reach full adult size and weight.

Older gorillas and chimpanzees, both male and female, often adopt orphaned siblings, allowing the youngsters to travel on their backs, sleep in their nests, and share their food. Even a silverback gorilla, the presumed father of most infants in his troop, has been known to take an orphan under his wing by sharing his nest and grooming the infant when he is frightened or startled.

When great apes are orphaned at the hands of humans—mainly by poachers, who kill the mothers and carry the traumatized infants to roadside markets or to middlemen who will smuggle them overseas—the results are even more devastating. Experts estimate that nine out of every ten infants taken from the wild do not survive the horrific ordeal. Those who do are psychologically scarred and physically weak.

Although conservation groups and primatological organizations are working with the local governments to curb poaching, ailing infant gorillas, chimpanzees, bonobos, and orangutans continue to surface in markets and back-alley showrooms across Africa and Asia, their terror-filled eyes pleading to return to the forest they will most likely never see again.

Attempts to return orphaned chimpanzees, bonobos, and gorillas to the wild after their confiscation from poachers have not been successful because of the youngsters' lack of survival know-how, skills they would have learned from their mothers. Although orangutan rehabilitation programs are under way in Borneo, the long-term success rate has yet to be determined.

above: *This chimpanzee mother and child are two of the fortunate few who live in the protected forest of Gombe National Park in Tanzania. The once-lush hillsides surrounding Gombe are now devoid of trees, vegetation, and wildlife as the surrounding villages continue to encroach farther into the unprotected forests.*

chapter iii

living in a complex society

In zoos, all four great ape groups seem to have similar social structures, diets, and habits. Although knowlegeable and caring zoos try to provide these intelligent residents with stimulating environments—ropes to climb, puzzles and toys to investigate, and a variety of edible treats throughout the day—no captive-born ape will know the joy of discovering a lush fruiting tree or the excitement of exploring a vast forest. Each day in the wild brings unknown challenges, new adventures, and, now, mounting dangers. Although, by necessity, all captive great apes appear to live similar lives in their zoo settings, in the wild the four great ape species have vastly different social structures, habits, and rules, each governed by habitat, food sources, and years of evolution.

Going Against the Crowd

The orangutan stands apart from the other great apes as the only species that spends the majority of its time alone. Orangutans have no groups, no cliques, no grooming circles. They travel alone, like large red ghosts, seeming to float across the forest canopy as they slowly and deliberately swing from tree to tree. The orangutan's solitary nature most likely evolved because of the wide distribution and shifting harvest of flowering and fruiting plants in its native habitat. The animals must remain on the move in order to find the ripened fruit. Because too little food is available at any one spot to support a group, they spread out, each orang with a wide expanse of forest overlapping the next orangutan's range. Amazingly, the orangutan carries with him a mental map of the forest layout—knowing from years of experience and training from his mother which fruiting trees will be ripe at what time, the location of the trees, and the most efficient route to take from any position in his forest range.

Their solitude is broken only occasionally—when males and females meet to breed, when adult males defend their range, and when females with their young gather to feed while their young ones romp. Although females do not seek each other's company, their accidental encounters within their overlapping ranges result in peaceful communal feeding and playing. Males, on the other hand, actively

opposite: *Gorilla groups are generally stable and quite peaceable. A silverback male and his troop of females and their young may travel together for several years.*

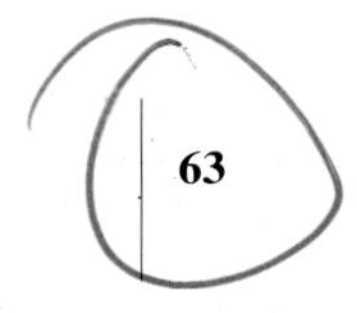

above: *A young Sumatran orang stares with shy curiosity at the imposing camera lens. Unlike the African great apes, orangutans are not social beings. They prefer instead to travel through the forests on their own, meeting up with the opposite sex for occasional consortships before moving back into self-imposed isolation.*

avoid each other's company, and the dominant male of any given range will emit a boisterous long call to warn other males of his presence.

A Social Life

In the African forest, where the fruits and edible plants are more abundant, each great ape species travels in groups, with individual community members often assisting each other in their search for food. Gorilla groups are the most stable of the African apes. The same adults, a silverback male and ten or eleven females and their young, travel together for months and sometimes years at a time. Each group has a home range, which varies in size from 1.5 to 11.5 square miles (3.9–29.8 sq km). And although there is considerable overlap in the ranges, little territorial defense exists among neighboring groups. Like orangutans, their large size somewhat restricts their mobility. Gorillas do not travel far, nor do they travel quickly, because their huge bodies need time to forage and digest. The gorilla's day is spent eating, resting, and eating again, before settling down in a comfortable nest of soft leaves and branches for the night.

Bonobos also lead relatively quiet lives. Like chimpanzees, they live in large communities that divide into small parties for daily travel. The composition of these parties changes from day to day—or even from hour to hour. This fission-fusion society, as primatologists call it, allows an individual to travel with his mother and younger siblings for part of the day, then respond to a far-off call from

left: *In bonobo society it is generally the females who have higher status and thus control access to highly prized foods. Though all the young may feed, adult males typically must wait until the females are satisfied before getting their share of the food.*

below: *Chimpanzees are highly social creatures, their fission-fusion society allowing them to associate with a number of different individuals within the larger group at any given time. Here, three chimps gather together for a quiet session of mutual grooming.*

adult males in another part of the range—perhaps to share in the rewards of a succulent fruit tree or to socialize with friends. Bonobo party sizes range from six to fifteen individuals within larger communities of as many as sixty. Unlike gorillas, bonobos and chimpanzees form groups with approximately equal numbers of males and females. After separating throughout the day, bonobo parties gather together at night, like hiking groups meeting at an agreed-upon campsite. Each individual and mothers with their infants bed in the trees on nests of bent branches and leaves, only to part again in the morning.

Chimpanzees also live in fission-fusion societies of up to 120 individuals, which break into smaller parties of three to six. The entire community's range can cover from four to twenty square miles (10.4–51.8 sq km), and individuals within the community generally stick to core areas within the larger range. Like bonobos, the smaller parties divide and recombine throughout the day.

what's for dinner

chimpanzees

The chimpanzee diet consists mostly of fruit, other vegetation, and some animal protein, including termites, driver ants, monkeys, and bushbabies. They also eat honey, eggs, and resin. Animal protein is obtained through collection (as for insects and eggs) and cooperative hunts, in which several males will team together to stalk, chase, and head off the prey.

bonobos

Bonobos eat mainly ripe fruits and the protein-rich pith from plants (leaf petioles and shoots of herbs). They also eat earthworms and caterpillars, but animal protein makes up only about 1 percent of their diet, and they have not been seen to hunt monkeys like their chimpanzee cousins.

gorillas

Gorillas are mainly leafeaters. Western lowland gorillas eat more fruit than eastern lowlands. They might also eat grubs when regular vegetation is scarce. Because the regular vegetation is generally a poor source of calories and protein, gorillas must spend the majority of their day foraging and feeding.

orangutans

About 60 percent of the orangutan diet is fruit; the other 40 percent includes young leaves and shoots, insects, flowers, bamboo, nuts, tree bark, and sometimes eggs and small vertebrates. They maintain a sophisticated mental map of the forest's bounty, keeping track of the location of all the trees and the seasons in which the fruits will be ripe and ready for consumption.

Follow the Leader

All groups need someone to take charge—whether in the boardroom, on the battlefield, or in the humid rain forests of Central Africa. The dominant characters in the great ape groups are usually males, yet the qualifications, rise to power, and method of leadership for this position vary greatly among the four species. Although females play a less obvious role in the hierarchical structure, their importance and influence on the social makeup should not be overlooked.

Life in a Gorilla Troop

The undisputed gorilla leader is the silverback, the patriarch of the community. Gorilla groups are sometimes referred to as harems because the silverback males surround themselves with females and their young, other males having left the group after adolescence in order to find a harem of their own. All the youngsters

in the group are likely sired by the silverback and they often shadow him throughout the day, as if searching for acceptance from their domineering dad.

Much of the gorilla troop's day is spent either resting or foraging—peaceful pursuits for which the silverback needn't spend much time exercising his mighty power. Yet he is always the focus of the group, and few movements are made without his knowledge or approval. When he does want the group's attention—if he is ready to move on to a new location or he's annoyed by someone's behavior—he beats his chest or the ground, tears up vegetation, or rushes toward the offender.

Although violence and aggression are rare among gorillas, fierce fights do break out between rival males—when an assertive blackback attempts to mate within his natal group before he takes off on his own or when a lone silverback happens upon an existing group. Lone silverbacks can be particularly threatening to an existing troop because they often kill the infants in the group—all sired by the silverback—in order to speed up a female's ability to mate again, this time with him.

When young female gorillas reach sexual maturity, at about the age of seven or eight, they are particularly attractive to the males in the troop, including the young blackbacks. During times of peace and tranquillity, the silverback remains the sole breeding male, taking his pick among the breeding females. Because the bond between the silverback and the females is strong, few female-to-female relationships develop.

opposite: *A silverback gorilla bares his massive canines in a gesture of dominance and threat, a behavior that rarely leads to actual violence in the peaceful gorilla society.*

below: *Silverbacks spend the majority of their day sitting quietly with their troop in the thick of the forest, sometimes interacting playfully with the young. When the silverback is ready to continue foraging, he alerts his troop with subtle verbal and nonverbal cues.*

An infant mountain gorilla joins others in an afternoon rest amid the thick vegetation of Rwanda's Virunga Mountains. Gorillas move from one foraging spot to the next throughout the day, following each eating session with a few moments of rest.

right: *Bonobos rely on affectionate physical contact to maintain their strong social bonds. While they are sometimes aggressive toward one another, they usually prefer to settle conflicts with a reassuring touch or sex rather than force.*

A Peaceful Matriarchy

In bonobo communities females form the core of the society and, unlike most mammalian societies, they often dominate the males. Instead of hierarchical ranking and dominance gained through size or aggression, certain female bonobos seem simply to earn the respect of their community. And older females maintain a higher status than those who are younger. The strongest attractions and bonds are among the females in each community and, to a lesser extent, between the sexes. Whereas food sharing may cause tension among male bonobos, females freely share their food with other females, and tend to have first dibs on a prized food source, forcing the males to wait in the shadows until they are offered a share of the bounty.

Female bonobos are certainly the "weaker" sex in size and weight, but their strong ties to each other and the importance that bonobo society places on sexual contact have piloted the female to the top of the bonobo pecking order. Indeed, males born to low-ranking females, or whose mothers are ill or no longer living, will never rise far in the male hierarchy. One eminent bonobo researcher, Professor Takayoshi Kano, who studies bonobos in the Congo's Wamba forest, describes the female and male bonobo hierarchy as a co-dominance. There may be one alpha male who can dominate the other males, but he is also the only male who can even attempt to threaten the alpha female. Because bonobos live in a fission-fusion society, no single individual leads the entire group. Instead, the particular composition of the small, ever-changing parties that make up the bonobo community dictates what will happen in any given situation. The bonobos' hierarchical structure continues to mystify and amaze the researchers who study them. But despite this puzzle of hierarchy—or perhaps because of it—bonobo gatherings of males and females are immensely peaceful and often highly amorous.

below: *In bonobo society, relationships among females tend to be the strongest. These friendships are developed and nurtured through food sharing, grooming—as in this photograph—and even female-to-female sexual contact.*

Political Males

While female relationships and hierarchies do play a role in the common chimpanzee society, it is the complex balance of male alliances and rivalries that determines the mood and composition of the community. Each community is ruled by an alpha male, who attained his position through carefully nurtured associations with potential allies, well-timed assaults on rivals, and the strength and skill to steady his forays up the hierarchical ladder. Because the large chimpanzee community divides into smaller parties throughout the day, the alpha male cannot have daily contact with or dominance over the entire group on a regular basis. But his status does afford him certain privileges, including first pick of the breeding females, first rights to the goods of the hunt, and the respect and attention of each member of the community whenever he is present.

Unlike gorilla silverbacks, however, the alpha male chimpanzee does not hold exclusive rights to all females. When a breeding female is in the vicinity, all males of high rank may follow her and initiate courtship at one point or another. Goodall counted old Flo, for example, involved in fifty sexual encounters in one day. If an alpha male, or another male of high rank, wants an exclusive relationship with a female in estrus, he must initiate a consortship—urging the female to run off with

right and opposite: *Chimpanzee society is governed by a complex hierarchy of males and females. Although the alpha male is the dominant figure in the larger community, the mood of each smaller gathering is dictated by the dominance and subordinance of each individual present.*

above: *Resting in the midmorning sun, this group of chimpanzees seems curious but undisturbed by the action taking place above them. Although fierce fights can break out among rival males, the majority of chimpanzee social gatherings are peaceful and amiable.*

him for a private getaway. He'll do this by waving branches, bristling his hair, and stretching his arm in the female's direction. If she is willing, the two will travel away from the party and out of the communication range of others in the group for as long as two to three days during the female's most fertile period. The purpose of the consortship is to eliminate other males' chances of mating with the female. Often the female will refuse the invitation for exclusivity, preferring instead to stay with the party and have her pick of interested males.

The presence of a breeding female is just one contributor to the tension-racked atmosphere and constantly shifting balance of power in the male chimpanzee community. New alliances are formed and authority tested on a daily basis. For this reason, the alpha male surrounds himself with a circle of supporters who will help defend him against potential usurpers. The supporters—or subordinates—know the alpha male will provide protection if they are ever under attack. Yet longtime

party loyalty is not part of the community makeup. A strong young male who supports the alpha as a young adult could very well attempt an overthrow if the opportunity presents itself later in life. Therein lies the difference between the bonobo and chimpanzee social makeup: bonobo communities revolve around the long-standing social bonds of the females, while chimpanzee communities are predicated on the volatile and opportunist coalitions formed by the males.

When an alpha male displays to express his annoyance or to remind those present of his authority, he'll bristle his hair to appear twice his normal size and charge screaming through the brush, breaking and hurling branches, slapping tree trunks and the ground, and hitting, pushing, and pulling anyone in his path of fury—including females and infants. During the display, the others shriek with fear and agitation as they scurry to get out of his way. Some huddle together, for apparent moral support, and mothers cower as they shelter their young.

above: *Hair bristled in anger, a male chimpanzee calls out threats and charges toward his intended victim. Alpha males intimidate subordinates and threaten rivals with impressive displays of anger and strength, rushing through the forest while slapping trees, breaking branches and limbs, and pushing or pulling any individuals in their way. This color-enhanced photograph depicts the drama of such displays.*

Following such a display the group will acknowledge the alpha male's supremacy by grooming him, offering their outstretched arms, or presenting their backs in a gesture of submission. Sometimes the mere threat of such an attack is sufficient to solicit submissive behavior. To issue such a threat an alpha male need only raise an arm, slap it in the air in the direction of the threatened individual, swagger bipedally, or throw rocks or other debris. If he's satisfied with the submissive response of the party, the attack goes no further.

Fights among adult males—either during an overthrow attempt or as a gang attack on a neighboring community—are much more violent than a display, and may even be deadly. When chimpanzees attack they will jump on the victim's back, pull hair, bite, drag their victims, and even slam them to the ground. Such gang attacks in defense of territory reveal a sophisticated level of teamwork, wherein group members may even hold an intruder down while others strike their blows.

Although Goodall had witnessed many territorial and hierarchical disputes among the males of the Gombe chimpanzees, none was as fierce or devastating as

left: *A chimpanzee screams with excitement and agitation, perhaps in response to an alpha male's display or an attack on a companion. Screams are generally high-pitched and carry far across the forest, sometimes bringing an ally to the caller's aid or comfort.*

the four-year war that took place in the early 1970s. Ten years after she began her research, she began to notice a division in the group—some animals remained in the north of the park, where she first encountered the chimpanzees, while others ventured farther south. As time went by, the relationship among the males of the two divided groups became more hostile, erupting into angry and noisy fights whenever each group's "border patrol" would cross paths. Slowly, the northern community began to systematically attack and kill their former companions—not only did they kill the rival males, but infants, females, and the elderly as well. In the end, the breakaway group in the south was completely annihilated. It was this discovery that furthered Goodall's and Leakey's understanding of the similarities between humans and chimpanzees, our closest genetic cousins. Not only do we share the ability to make and use tools, to form complex social groups, and to feel and show emotions—chimpanzees, like humans, are capable of sophisticated and premeditated warfare.

above: *The fully mature male orangutan possesses large cheek pads—once he has developed these cheek pads he will be able to attract receptive females.*

opposite: *Only young orangutans seem to take pleasure in each other's company. These younsters seize the opportunity for play when their mothers gather at a fruiting tree.*

An Understated Hierarchy

Even among the largely solitary orangutans there exists a dominant male and a code of conduct that must be followed in his presence. Although male orangutans are even less social than the females, who may gather on occasion with their young at a fruiting tree, the males maintain a system of hierarchy by recognizing each other's long calls. The calls begin with low rumblings, grow into loud groans, and end with booming roars that bellow throughout the forest and can be heard over a kilometer away. Because each male has a distinctive long call, he communicates to others his location and stakes his claim on a particular area—or a particular female. Young males do not emit long calls, and larger, more dominant males call more often than other adult males, causing subadult males to move quietly out of range.

When male orangs meet face to face, their encounters can be aggressive. They inflate their pouches and charge at each other through the trees, breaking branches and hurling them to the forest floor. If the threat is not enough, they grab, wrestle, and bite each other's cheekpads or ears. Natural selection is evident in this primitive rise to power. Because it is brute strength—not alliances or cunning—that ushers male orangs to dominance, the larger males almost always get the girl, and thus go on to breed.

Even though a dominant male has won the right to mate with females in his area, the females are not always agreeable. Many researchers have witnessed male orangutans forcing copulation with unwilling females. When the mating is mutually acceptable, the couple will steal away for day- or week-long consortships during which intercourse is their top priority. But when the female is not willing, the male orangutan will simply pin her down, through screams of protest, before rambling away to continue his solitary existence. Interestingly, females rarely become pregnant as a result of forced copulation.

Dominant males cease their sexual activity at about the age of thirty, but they continue to defend their long-protected territory. This leads scientists to speculate that the males are maintaining dominance over the area until their male offspring are able to take over the space.

Keeping the Peace

Of all the great apes, the bonobo holds the title as the most peaceful. They maintain this peaceful atmosphere in the most provocative of manners: through sexual and erotic contact. And because this contact occurs just as often between partners of the same sex as the opposite sex, it is obviously not for the sole purpose of reproduction. In bonobo society, sexual contact eases tensions, resolves conflicts, and is even used to trade for food. Female bonobos with sexual swellings often engage in intercourse with a male before stealing his prized fruit. New females who have transferred from another group will use sexual contact with the resident females to ease their way into the new society. And males will gently massage each other sexually after a fight—perhaps as an effort to reassure one another that the disagreement is indeed over.

Sexual contact also appears to dampen heightened excitement or tension over food. When a party of bonobos encounters a succulent fruiting tree they often partake in sexual intercourse and contact—even between members of the same sex—before beginning to feed. Indeed, bonobos are the only great ape believed to mate just for sheer entertainment.

Chimpanzees also use physical contact as a means to ease tensions, but it comes in the form of grooming, touching, embracing, or kissing. Their need for physical contact in times of stress most likely stems from early childhood, when a mother's embrace or touch was enough to calm fears. Social grooming is the most common form of peaceful interaction among adult chimpanzees. Grooming clusters—something like our own shoulder massage "trains"—can involve as many as twelve individuals, all sitting together in blissful harmony as they sift through each

opposite: *Young orangutan males like this one in Borneo's Tanjung Puting Park do not develop their cheek pads or throat pouches until they are at least fifteen years old. Yet even then they will not emit the booming long call of an adult male if there is a dominant male already living in the same area.*

right: *Bonobos, like their chimpanzee cousins, use physical contact to form and strengthen bonds, especially among females. Another strong relationship in bonobo society is between mother and son, a friendship that often lasts a lifetime.*

other's hair, picking out dirt and debris and engendering a sense of calm and harmony among the participants.

As with many activities in chimpanzee society, politics plays a part in the daily grooming activities. Lower-ranking male chimpanzees are more likely to groom those with a higher status than the other way around—presumably in an attempt to gain a valuable ally. Rival males with tense relations will groom each other more frequently than they will other males, perhaps to keep the other at bay.

Following a male chimpanzee's aggressive display, peace and order is restored when group members show submission and the aggressor offers reassurance that all is now well. Submissive gestures include presenting—or turning one's rump to the aggressor—reaching out with upturned palm, and embracing, often accompanied by squeaks, whimpers, or pathetic squeals. To reassure a submissive subordinate, the aggressor responds with touching, patting, or grooming. Sometimes a third party, one not involved in the initial conflict, offers assurance or comfort to the frightened victim by patting him on the back or reaching out a hand of appeasement, a true illustration of the chimpanzee's capacity for empathy and understanding.

As with humans, chimpanzees and bonobos also kiss to show affection or end an argument. An aggressor may kiss the hand of a recent victim to show appeasement or reassurance, or two individuals may engage in a mouth-to-mouth kiss to signify the end of the conflict. Two individuals often kiss when meeting each other after an absence.

Just as we demand that others look us in the eyes to prove that what they say is true, chimpanzees and bonobos seek eye contact when resolving conflict or forming alliances. To look away is an act of distrust and deceit.

Gorillas maintain peace in a less demonstrative way. Their calm demeanor and the group's natural acceptance of the silverback's dominance set the tone for quiet gatherings with few adult interactions. When mutual grooming occurs, it is frequently between a mother and her infant or a female and the silverback. Social grooming among adult females is quite rare. In fact, the silverback's dominance is so well-accepted—both by subadult males and by the females in his troop—that internal squabbles are virtually nonexistent. Unlike chimpanzees, bonobos, and humans, gorillas avoid eye contact, viewing it as a sign of aggression or challenge.

Grooming and other forms of physical contact are also rare among orangutans. But even though they do not embrace, share food, or groom each other, orangutans—especially females and their young—do maintain a distant relationship. They know the other individuals that share their territory and they can form enmities and loose alliances. Although females seldom interact in the same area, they show signs of recognition and acceptance, as if acting out an elaborately choreographed pantomime, while their young ones play nearby. In short, the orangutans' best method of peacekeeping is simply to avoid conflict by monitoring the whereabouts of adversaries. And although a mother orangutan is tender, nurturing, and affectionate with her infant, the young orangutan does not seem to carry this need for physical reassurance and interaction into adulthood.

chapter iv

the great ape mind

Throughout time humans have created definitions for ourselves that separate us from the animals—toolmaker, communicator, thinker, and bearer of emotions. Without question, humans stand apart from the animal world for a multitude of reasons. But our continuing research of the animal kingdom reveals that humans are not as unique as we once believed. Great apes communicate; reason; feel joy, pain, and grief; and have empathy for others.

Opening the Window

Intelligence in nonhuman animals can be difficult to measure. How can we determine whether another species processes information efficiently without even knowing how their minds process information at all? An orangutan's sophisticated mental map of her range and memory of the seasons of each fruiting tree is an impressive feat, difficult for a human to duplicate. But do we judge her intelligence based on this capacity in the wild, or on a human-devised experiment in a laboratory?

When researchers first realized that the great apes mirror our own species in behavior and chemistry, they pulled young specimens from the wild and began a barrage of experiments to test for cognition, memory, and problem-solving. The most noteworthy of these early studies were conducted by German psychologist Wolfgang Köhler on the island of Tenerife in the Canaries and by Robert Yerkes, an American researcher working at the Yale Laboratories of Primate Biology.

Köhler began his studies in 1914 with a group of nine chimpanzees. He noted that the "higher apes" are "nearer to man than to the other ape species" and thus began positing questions about their behavior, saying "these beings show so many human traits in their 'everyday' behavior . . . the question naturally arises whether they do not behave with intelligence and insight under conditions which require such behavior."

Köhler set about to answer this question by placing bananas or other fruits within the chimpanzees' sight but out of reach, requiring the use and manufacture of tools to obtain the rewards. The chimps performed the tasks admirably—stacking boxes, joining sticks, and reaching with poles to seize the prized fruit—leading

opposite: *A bonobo looks skyward with intense concentration. Despite the bonobo's obvious intelligence and their similarities to chimpanzees in diet and habitat, they have not been seen using tools in the wild.*

binti jua, b. 1988

gorilla

Gorillas are the gentle giants of the African jungle. They seldom fight, quarrel, raise their voices, or cause a scene. Gorilla mothers are attentive, thoughtful, nurturing, and protective. We witness their peaceful demeanor in zoological parks and nature shows again and again, but it takes an extraordinary event to truly impress upon us the emotional depths of this remarkable species. This extraordinary event occurred in August 1996.

Binti Jua (Kiswahili for "Daughter of Sunshine") is a western lowland gorilla who lives in a harem at Illinois' Brookfield Zoo. On this warm summer day, she sat with the others in the outdoor enclosure and nursed her seventeen-month-old infant, Koola, as hundreds of laughing, screaming, and curious onlookers paused, then continued past the exhibit on their journey through the zoo. It was a day like any other until one three-year-old boy dared to get a closer look. With no supervising adult in sight, the boy climbed the small fence that surrounds the sunken gorilla exhibit, lost his balance, and fell eighteen feet (5.5m) to the concrete floor.

To the horror of all who looked on, the young boy lay unconscious in a den full of startled three-hundred-pound (135kg) gorillas. Binti acted without hesitation. With onlookers screaming in terror at the potential outcome and young Koola still clinging to her breast, Binti approached the boy and gently lifted his arm, as if looking for signs of life. She then gathered him into her arms, just as she held her own infant Koola, and looked up into the crowd. When Alpha, a larger female, came near, Binti uttered a guttural warning that stopped Alpha in her tracks. Finally, after a moment that could have been filled with nothing less than reasoning, Binti carried the injured boy to a corner of the exhibit and laid him gently at the keeper's door, where he was rescued just moments later.

Thankfully, the young boy suffered only minor injuries and was active again in just a few short days. Meanwhile, Binti made headlines the world over. One witness to the event told Life *magazine, "I can't help but think there's a message in this. She didn't hesitate to help. If this animal that's supposed to be below us can be this way, why can't we?"*

Why indeed?

Köhler to conclude in his 1929 book *The Mentality of Apes,* "the chimpanzees manifest intelligent behavior of the general kind familiar in human beings."

Less publicized but equally impressive were Köhler's sensitive and insightful observations of the captive chimpanzees' behavior outside the testing yard—perhaps a better yardstick by which to measure their mental abilities and intelligence. Köhler realized, even in 1914, how dependent these captive chimpanzees were on each other's company, how they would defend each other against presumed attacks or dangers, and how each chimpanzee was a unique individual, with his or her own disposition and personality.

Shortly after Köhler's work began, psychobiologist Robert Yerkes began a study of chimpanzee behavior at the Yale Laboratories. Yerkes later acquired more chimpanzees, as well as gorillas and orangutans, to form the Yerkes Primate Laboratory in Orange, Florida, and finally the Yerkes Regional Primate Research Center in Atlanta, Georgia, which is still in operation today. His early conclusions on the mental capacities of chimpanzees echoed those of Köhler. Yerkes went on to rank the gorilla as superior to the chimpanzee and orangutan in attention, memory, and imagination—a distinction that has yet to be confirmed or disproved.

below: *Facial expressions of all the great apes mirror those that humans show when feeling sadness, fear, or joy.*

A contemporary of Köhler and Yerkes was a young Russian psychologist named Nadie Kohts, who "adopted" a year-and-a-half-old chimpanzee in 1916. She raised the young ape as her own child, presenting him with hundreds of cognitive tests and problems until he was four years old. A few years later she repeated the process with her own son during his first four years.

These early studies opened a window into our understanding of the mental capacities of the great apes, emphasizing the apes' ability to solve a multitude of puzzles and problems. Turning the tables, one wonders how you or I would fare if pulled from our lives of automobiles, telephones, and grocery stores and placed in the dense undergrowth of an equatorial rain forest. Would we know, without prompting, how best to pull the biting termites from their protected mounds, which of the thousands of plants are edible and tasty, or how to communicate to our primate cousins that we mean no harm?

Studies on great apes' intelligence continue around the globe, and range from work with Ai, the female chimpanzee in Japan who discriminates characters on a computer-controlled device, to the orangutans in the National Zoo's Think Tank, who communicate with their keepers through cognitive tests involving tool use and language.

left: *Chimpanzees use a variety of postures, facial expressions, and hand gestures to communicate nonverbal messages to their companions, leading Robert Yerkes to postulate that, if they can't speak the human language, perhaps they can communicate with humans using their hands.*

opposite: *A lowland gorilla strikes a studious pose, leading onlookers to wonder what complex thoughts are on his mind. When Robert Yerkes began his studies in the early 1920s, he ranked gorillas superior among all great apes in attention, memory, and imagination.*

Redefining Humans

A greater understanding of the great apes' complex minds, intricate thought processes, and remarkable capabilities began in earnest when researchers allowed the apes to stay in the wild, studying their responses and solutions to natural dilemmas rather than those contrived by humans.

Not long after Goodall began her study of the chimpanzees of Gombe, she saw a chimpanzee named David Greybeard pick and strip a blade of grass, ridding it of offshoots and debris to make it smooth and straight. He carefully inserted the blade into a termite mound through a hole he'd dug with his fingernail, waited for the termites to latch onto the blade with their mandibles, and just as carefully moved the blade from the hole to his mouth, eating the termites off with great precision.

right: *Chimpanzees in West Africa use rocks and logs as makeshift hammer and anvil tools to break nuts, a behavior that has yet to be seen among wild chimpanzees in other parts of Africa. Interestingly, this orphaned chimpanzee resides at a sanctuary in East Africa. Has she imported this behavior from her life in the wilds, and will other sanctuary residents soon adopt it?*

When Goodall reported to Leakey her finding that chimpanzees not only use tools but modify objects to make tools—a trait thought to be exclusively human—Leakey responded, "We must now redefine tool, redefine Man, or accept chimpanzees as humans."

The finding that chimpanzees make and use tools in the wild opened a debate about the definition of intelligence and inspired a hunger to learn just how far great apes could go in adapting their own surroundings.

Chimpanzees use branches, vines, and twigs, in addition to blades of grass, to fish for termites, each time taking great care to find one of the right size and strength and then manipulating it, stripping it of leaves or branches, to perfect the tool for the job. Large sticks and branches are used by chimpanzees to break into bees' nests and investigate animal holes or dens.

When water collects in tight spots chimps can't reach with their mouths, they chew leaves to a mushy pulp and use it as a sponge to absorb the water. Then they

above: *A group of chimpanzees gathers excitedly around a termite mound to fish out the succulent insects with twigs, sticks, and grasses. Termite fishing requires know-how in choosing and manipulating just the right tool, as well as composure and coordination in removing and eating the termites.*

Studies with chimpanzees have shown them to have imaginations, a trait previously believed to be exclusively human. For intance, a chimpanzee may pretend that an inanimate object, such as a teddy bear, is alive or may treat a purse as if it were a shoe, a type of imagination known to psychologists as "substitution."

suck the water from the sponge, repeating the process until their thirst has been quenched. Leaves are also used as primitive napkins when chimpanzees become soiled or sticky from the juice of a fruit; they also use leaves to dab at blood when injured.

Tool use methodology is not universal among all chimpanzee populations. The termite-fishing technique is common in the Gombe community in northern Tanzania, in the Mahale Mountains south of Gombe, and in the Mount Asserik region of Senegal, among other places. But the chimpanzees in Mbini and West Cameroon use larger sticks to fish for termites. And in Guinea, they pound the sticks up and down rather than inserting them carefully into the mound to get the catch.

right: *A young chimpanzee pounds a stick into the ground to investigate an animal den. Although this type of tool use may begin as a simple game during childhood, it is a skill the chimp will continue to use and perfect well into adulthood.*

above: *The community in which a chimpanzee is raised will affect the ways he'll use tools and perhaps other behaviors as well. This type of learning, passed down from generation to generation, is an example of primitive culture.*

Chimpanzees in West Africa use stones and large sticks as a hammer and anvil to open oil-nut palm seeds or other hard nuts, a technique never seen in the Eastern communities, even though the same seeds are abundant throughout chimpanzee habitats. These differences illustrate the cultural traditions and learned behaviors of the chimpanzee populations. It is impossible to know when the first chimpanzee discovered that he could open a nut by crushing it between two stones, but it is clear that the behavior was imitated and became a common practice among his community and for each generation that followed.

When a chimpanzee named Mike in the Gombe community began to use empty kerosene cans as a scare tactic in his display, throwing and banging the "tools" he'd found in Goodall's camp to create a thunderous stir among higher-ranking males, he was adapting his surroundings in a novel way to perform a premeditated act. At one point during his rise to power, he walked into Goodall's camp unagitated, carried the cans toward the other males, and waited for just the right moment to commence his display.

Tool use on its own does not reflect a great intelligence or higher mental function. Even some insects are known to use objects to accomplish a task. It is the chimpanzee's manipulation of objects to make them suitable for a particular task that points to the great ape's remarkable mind. As Goodall points out in *Patterns of Behavior*, "The chimpanzee, with his advanced understanding of the relations between things, can modify objects to make them suitable for a particular purpose. . . . He can pick up, even prepare, an object that he will subsequently use as a tool

at a location that may be quite out of sight. Most important of all, he can use an object as a tool to solve a completely novel problem."

Orangutans have long been considered the master tool user and escape artist in captivity, confounding even the most careful caretakers with sophisticated breakouts. Benjamin Beck, project director at the National Zoo, described orangutans as "reflective, contemplative, insightful, and deliberative" in an article in *Smithsonian* magazine. According to Beck, if a zookeeper left a screwdriver within a chimpanzee's reach, the chimpanzee might fiddle with it for a while and then move on to something else. An orangutan, by contrast, would pretend not to notice the screwdriver until the keeper left, then he would use it to dismantle his cage and escape. It is only within the past five years that researchers discovered orangutans manufacturing and using tools in the wild—breaking and stripping branches to extract honey from nests.

opposite and below: *Orangutans lead complex lives in the wild—from maintaining and following a mental map of the fruiting forest trees to knowing how to reach the edible part of the strychnos fruit. But we have only recently begun to discover the orangutan's use of branches to extract honey—a form of tool use in the wild.*

indah, b. 1980

orangutan

Visitors to Washington, D.C.'s National Zoo straining their necks to look up in the sky aren't looking at passing airplanes, low-flying birds, or nearby skyscrapers. More than likely they are watching the orangutans in their daily commute between "home," the Ape House where they sleep and relax, and the zoo's Think Tank, where they interact with computers and caretakers while providing visitors with a glimpse into the orangutan mind.

The two buildings are linked by a series of eight 45-foot-high (14m) O-Line towers connected by steel cables that hover over the zoo's public sidewalks—providing zoogoers with a more realistic perspective on locomotion. The platforms are adorned with electrified skirts to keep the orangs from descending into the zoo, but during daylight hours—when the keepers are present—they're free to go from Think Tank to Ape House as they choose. Indah, one of the star students in the program, has chosen to make the Think Tank her permanent home. And her older brother, Azy, makes frequent visits as well.

The goals of the Think Tank are many-sided. The entire exhibit compels visitors to examine animals' thought processes and compare them to our own. Through interactive games and well-designed displays, we look at instinct versus thinking in animals from salmon to beavers—and, of course, orangutans. It is also an attempt to show people the intelligence and sensitivity of the orangutans, in the hope that more individuals will join the fight to save the vanishing species.

With Indah and Azy, visitors are able to watch true brain power in action. In full view of the public, Project Director Rob Shumaker shows Indah a particular fruit and waits for her to select the appropriate symbol on the computer screen and press the "send" key. When she answers correctly, as she invariably does, she is given the item she correctly labeled. Indah understands the symbols for "banana," "apple," and "grape," among others, and her caretakers expect that she will continue to add new symbols to her list. Other demonstrations illustrate the orangutans' reasoning and problem-solving skills. Rob Shumaker and assistants put on elaborate performances to test their theories. In one case, Shumaker places food under one of two cups out of Indah's reach. He leaves the room and is replaced by two other zoo employees who did not see where the food was hidden. The two employees stand with their backs to Indah, waiting for instruction. The purpose of the exercise is to see if Indah can model the employees' visual perception by letting them know under which cup the food is hidden. Although orangutans are not known for extensive tool use in the wild, the project also focuses on their keen tool-using abilities in captive settings, including dragging treats from out-of-reach corners or small openings using bamboo rods.

Orangutans in the wild face a multitude of daily challenges—locating the nearest fruiting tree, prying open prickly durian fruit, maneuvering through the unpredictable upper branches in trees. Although Indah's mental challenges in the Think Tank may pale in comparison to a wild orangutan's fight for survival, her keen intelligence and quick mind make her an ambassador not only for her own species but for all animals whose minds we have yet to fully appreciate.

Bonobos, by contrast, have not been found to use tools in the wild, nor do gorillas, although both species use tools regularly in captivity. The abundance of food in the wild gorilla's diet may be the reason they haven't resorted to complicated methods of food retrieval. But the bonobo's lack of tool use in the wild remains one of the many mysteries of this great ape.

left: *Though captive bonobos are adroit tool-users, bonobos in the wild have not been observed using tools. Some scientists remind us, though, that the discovery that orangutans use tools in the wild is relatively recent—perhaps there is more to be learned from other bonobo populations.*

Talking with the Animals

Where does a 400-pound (182kg) gorilla sleep? The answer to the joke would be "anywhere she wants," but in Woodside, California, you can simply ask the gorilla yourself. Koko is just one of a growing number of captive great apes who have been taught to communicate with humans using the hand signals of American Sign Language (ASL).

right: *Some gorillas who have been taught sign language use it not only to communicate with their human companions but will also sign to themselves, just as humans occasionally talk to themselves.*

roger fouts, b. 1943

"This laboratory is Washoe's home. You are guests in her house . . . You are free to walk out of here anytime you like, but Washoe's family can never leave. Your job is simple: to make their lives as pleasant, as social, and as interesting as possible."

This speech is prepared for visitors to the Chimpanzee and Human Communication Institute in Ellensburg, Washington, home of Washoe and other chimpanzees—and home-away-from-home for the institute's codirectors Roger Fouts and his wife, Debbi.

Roger and Washoe are kindred spirits. Neither one planned to be the center of an intensive study on human and chimpanzee communication that placed them in the middle of a debate over the rights of chimpanzees. Roger had intended to be a child psychologist and Washoe, born in the wilds of Africa, would have lived her life in the rain forest. But their fates collided when, in the early 1960s, the United States stepped up its space program. Hundreds of infant chimpanzees were captured in Africa and imported to Holloman Air Force base in New Mexico for intensive testing for a journey into space. One of these unfortunate captives was a young female, later named Washoe.

By the time this infant arrived in America, the space program no longer needed chimpanzees. The new recruits were being sent to medical laboratories around the country for use in medical experiments. This youngster narrowly missed a horrific fate. Instead, she was taken home by Drs. Allen and Beatrix Gardner, who named her Washoe and set out to teach her sign language while surrounding her with a comfortable home and human companions as her family and friends.

Meanwhile Roger, searching for an open door to launch his career in psychology, uprooted his young family from California and settled at the University of Nevada, in Washoe County. He went in search of a half-time research assistantship with a husband and wife team who had a grant to raise a chimpanzee. And this is when Roger first met Washoe.

Initially, Roger worked with Washoe as the Gardners' assistant, helping as the chimp mastered thirty-four signs within two years and eighty-five signs one year later. She even signed to herself, like a self-absorbed child in an imaginary world, and combined signs to create names for words she didn't know: "dirty good" for toilet and "open food drink" for refrigerator.

In 1970, after Roger had attained his goal of a doctorate, the research took a different turn. The Gardners stepped out and Roger, with his wife Debbi and their two children, took a leading role in Project Washoe and in the future and well being of the project's star. The communication study continued and was expanded to include more chimpanzees. The study began to explore the variation in ASL use among the chimpanzees and their capacity to acquire signs from each other. In 1980 the growing family of humans and chimpanzees made another move—this time to the Pacific Northwest to establish the Chimpanzee and Human Communication Institute at Central Washington University in Ellensburg. Studies at the institute focus on communication among the chimpanzees and Washoe's own use of ASL around her adopted son Loulis.

Washoe's ability—and willingness—to communicate with humans in a human-devised language has opened up a world of wonder and, at times, disbelief. Few can imagine what it's like to exchange thoughts with a chimpanzee, and Roger never takes the privilege for granted. His commitment to Washoe and her family and his gratitude for all she has taught him over the past thirty years are strong and unwavering. With the insight he's been given by a wild-caught chimpanzee named Washoe, Roger Fouts has launched a personal campaign for the rights of chimpanzees in captivity, lobbying for humane treatment and enriching and stimulating environments—for the chimpanzees who can't speak for themselves.

above: *There are few animals that can produce the noise level of a group of excited chimpanzees.*

opposite: *A chimpanzee calls out to his companions.*

previous pages: *An infant mountain gorilla cries out for his mother.*

Attempts to teach great apes to speak commenced soon after Köhler and Yerkes announced to the world the results of the intelligence tests they'd conducted. If these animals can think almost like humans, people reasoned, can they not speak like us as well? The first tests were resounding failures. In the 1940s researchers adopted a female chimpanzee named Viki to raise like a human child, encouraging the young ape, as parents do, to pick up words and imitate sounds. But despite Viki's intelligence and her surrogate parents' encouragement, the young chimpanzee could utter only four hoarse words: *mama, papa, up,* and *cup.*

In 1925 Yerkes had predicted the next step in human–ape communication. "I am inclined to conclude that the great apes have plenty to talk about, but no gift for the

use of sounds to represent individual … feelings or ideas. Perhaps," he surmised, "they can be taught to use their fingers." And that is exactly what happened.

The trend of teaching sign language to apes began in 1966 with a wild-born chimpanzee named Washoe. She, like Viki, lived the life of a human child (except that her bedroom was in a trailer rather than in the main house). From an early age Washoe was immersed in the language of the deaf by a husband-and-wife team of researchers, Allen and Beatrix Gardner, and later by Dr. Roger Fouts and his wife, Debbi.

The researchers began to realize that Washoe was not simply imitating the hand gestures used by her caretakers when she invented her own signs: "open food

right: *Indah, an orangutan, chooses the correct symbol for "apple" on the computer touch screen when an image of an apple is flashed on the projection screen in front of her.*

above: *A bonobo reaches out in a gesture of friendship and reconciliation.*

drink" for refrigerator and "dirty good" for her toilet seat. She even signed to herself, like a child deep in imaginary play: "quiet" when sneaking to a part of the yard that was off limits and "hurry" as she hastened to the "dirty good." Even more amazing was the discovery that when Washoe was joined with other chimpanzees who knew sign language, they conversed with each other in their strange new language—entirely independent of any caretaker's participation or encouragement. And when Washoe adopted an infant male chimpanzee, the Fouts ceased nearly all signing in his presence to see whether he would pick up the language from his adoptive mother and her companions. As Roger Fouts reports in *Next of Kin: What Chimpanzees Have Taught Me About Who We Are*, after eighteen months with the signing chimpanzees, the young adoptee knew and used nearly two dozen signs. Washoe lives with four other signing chimpanzees at the Chimpanzee and Human Communication Institute, run by Dr. Fouts and his wife, Debbi, at Central Washington University in Ellensburg, Washington.

The gorilla Koko's adventure in the world of human communication began in 1972 at the San Francisco Zoo, where the one-year-old female gorilla was taught ASL by researcher Francine Patterson. By the time she was five and a half years old, Koko had mastered some two hundred fifty ASL signs, and now has a working vocabulary of more than five hundred signs. Koko and her male gorilla companion, who also signs, live at Gorilla Foundation, run by Patterson and her assistants, outside Woodside, California.

Orangutans have also been taught to use sign language, though the studies have not had the longevity or the notoriety of the experiments with Washoe and Koko. Chantek, a male orangutan who lives at the Yerkes Primate Center in Atlanta, mastered fifty-three signs after only eighteen months of training. His tutors and caretakers describe him as "curious and inventive," frequently offering his hands to be molded when he wanted to know the sign for an object and even inventing his own signs by combining two words he knew, such as "eye drink" for the contact lens solution used by a caretaker. Chantek's trainers also described his signing as "slower and more articulate" than that of chimpanzees who signed, concluding that this "insightful cognitive style was evidence of the orangutan's linguistic superiority over the African apes."

Despite these successful and ongoing studies, the great apes' ability to "speak" sign language is bitterly debated. Are they really communicating their thoughts, needs, and emotions? Or are they merely reacting to cues from the researchers or to conditioning? Fouts's and other notable scientists' reputations for integrity and solid scientific research should help to dispel doubts about these apes' abilities.

To further solidify the notion that great apes have the capacity to think abstractly, form sentences, and express emotions, one need only meet Kanzi, a bonobo who lives with a colony of other bonobos and chimpanzees at the Georgia State University Language Research Center near Atlanta. What makes Kanzi stand out among other "talking apes" is his equally impressive computer literacy. Kanzi speaks with his tutor, psychologist Sue Savage-Rumbaugh, by punching keys on a special keyboard or pointing to symbols on a board. His grammatical comprehension is comparable to that of a two-and-a-half-year-old human child. Because Kanzi's study centers around stricter and more measurable scientific guidelines, his—and Savage-Rumbaugh's—achievements have further renewed interest in the intelligence of our closest animal cousins.

right: *The bonobo Kanzi uses a computer with special symbols to communicate. Researchers working with Kanzi often make use of headphones so that the assistant present does not inadvertantly give the bonobo clues to the correct answers.*

kanzi, b. 1981

bonobo

His favorite movies are Tarzan, Iceman, *and* Quest for Fire. *He gets a kick out of people wearing gorilla suits. And he spends a great deal of time tapping his fingers on a computer keyboard. His name is Kanzi, and he is a bonobo who has taken the research community by storm.*

Kanzi fell into the spotlight quite by accident. He was born to Matata, a wild-caught bonobo involved in a study at the Georgia State University Language Research Center. Psychologist Sue Savage-Rumbaugh tried to teach Matata a system of abstract visual symbols presented on a keyboard. Although Matata was adept at communicating her needs and desires to Savage-Rumbaugh through gestures and bonobo-like vocalizations, she paid little attention to the lexigrams. Little did Savage-Rumbaugh realize that two-and-a-half-year-old Kanzi was eavesdropping on his mother's tutoring sessions all along. He seemed to understand much of the spoken language used around him, and he used the lexigrams to obtain what he wanted. He might press "bite" if he wanted a bite of something Savage-Rumbaugh was eating—or "chase apple" right before he grabbed an apple and darted away, expecting to be chased. By the time he was six, Kanzi had a vocabulary of two hundred symbols.

Involved in a study that probes the origin of human language, Savage-Rumbaugh came to a realization: human children do not learn language in formal, classlike settings, as she'd been attempting with Matata; they learn language by observing, listening, and experimenting. And this, she decided, would be the basis for Kanzi's training.

But because Kanzi is a bonobo, a sterile lab or even a private home did not seem appropriate venues for the study. Wild bonobos spend their days wandering through the forest, Savage-Rumbaugh reasoned, so the best atmosphere for continued learning would be the fifty-five acres (22ha) of forest behind the Language Research Center. The experiment soon continued with Kanzi's sister, Panbanisha. Each day Kanzi and Panbanisha join Savage-Rumbaugh and her research assistants for a walk among the trees—where the researchers converse and use the keyboard to discuss where they go, what food they are going to eat, and what they will do next. And the bonobos join in. In one case a researcher in the yard with Panbanisha heard what she thought was a squirrel, and typed "there's a squirrel" on the portable keyboard. Panbanisha responded by typing "dog." Soon three dogs appeared in a far corner of the yard.

Although it is interesting enough that such an exchange can take place between humans and nonhumans, Kanzi helped to push the envelope by responding appropriately to words used in new combinations. When Savage-Rumbaugh typed "Go get the ball outside," Kanzi bypassed the ball that was inside on his way to retrieve the ball Savage-Rumbaugh had requested.

When teenaged Kanzi was placed head-to-head with a two-year-old girl in a language comprehension test of six hundred sentences, they each had their strong and weak points, but in the end their scores were pretty much the same. But Kanzi isn't concerned with test scores, or language studies, or whether his remarkable accomplishments are sending shockwaves through the language research community. He knows that when he types "Fire TV" he gets to watch Quest for Fire. *And that, for now, seems to be enough for him—until he tells us otherwise.*

Communication in Nature

Ongoing studies on the meaning of great apes' vocalizations in the wild, the nuances of their facial expressions, and the purpose of their many postures continue to entice primatologists, anthropologists, and linguists alike.

Although the reclusive orangutan is the least vocal of the great apes, the Indonesian rain forest echoes on occasion with the bellowing long call of the dominant orangutan male. The call begins as a soft grumble, builds to a roaring crescendo, and terminates with more grumbles and a final sigh. According to legend, the male is pining away for his human bride, who escaped from his treetop nest. In fact, the call seems to have a multitude of functions. Subordinate or rival males move away from the caller, and some females head toward the call. Perhaps, then, the lone crier is signaling to his forest mates where he is—warning other males to stay away and urging females to join him.

Other orang vocalizations that may break the silence of the forest include squeaks, moans, barks, and screams—usually emitted by youngsters during play or frightening situations. The solitary nature of the orangutan leads one to believe

right: *The roaring long call of the adult male orangutan booms through the forest, warning other males to keep their distance and inviting females to join him.*

left: *Gorillas are the quietest of the great apes, relying mainly on soft grunts and nonverbal signals to communicate with each other.*

that gestures and other nonverbal communication would be unnecessary, but when encounters do take place, it is the nonverbal clues that allow orangutans to convey their intentions. When two males meet they stare each other down, inflate their pouches, and purse their lips together or open their mouths and gape widely in an exaggerated yawn to display their teeth, all with rigid posture to augment their size. Frightened or submissive orangs protrude their lips, often accompanying this expression with soft whimpers. And orangutans engaged in play dangle their lower lips and draw the corners of their mouths back in what can only be described as an orangutan smile.

Despite their size and undeserved reputation for fierceness, gorillas are runners-up as the quietest great apes. When they do find the need to vocalize their feelings, their roars are the most explosive sounds in nature. Silverbacks give forth alarming roars and screams when startled by approaching buffalo, other silverbacks, or humans. When mildly agitated they emit soft growls, and females pant

right: *A silverback gorilla generally need only rise and begin walking to communicate to his troop that he's ready to move. When he does feel the need to command more attention, he'll emit soft growls or barks.*

deeply when involved in minor quarrels with other group members. Juveniles screech, scream, or wail when frightened or unhappy.

On less distressing occasions, silverbacks emit soft piglike grunts or barks to gather the group together or encourage them to move on—to which group members may grunt in response, signaling "we hear you and we're on our way." Young gorillas involved in play actually chuckle. In all, Fossey encountered up to sixteen gorilla vocalizations, each occurring in a specific and documented situation and often resulting in a specific reaction from the gorillas nearby.

The most recognizable form of nonverbal communication among gorillas is chest beating, a performance often accompanied by beating of the thighs, shoulders, and nearby tree trunks. Researchers calculate that a gorilla in full display can beat his chest at a rate of ten slaps per second. If a silverback is protecting his group—or if a lone silverback is attempting to take over his troop—he beats his chest, shakes trees, stomps the ground, and emits staccato hoots that end in a loud, rumbling roar. During the hoot series the male may place a small leaf or twig between his lips—a phenomenon whose purpose still confounds primatologists.

Indeed, the meaning behind many of the facial expressions and postures of wild gorillas remains the subject of research for primatologists in the field. What is known, however, is that several of the expressions and communicative postures mirror those of common chimpanzees—and, at times, of human beings—pointing once again to our common ties.

Chimpanzees possess a large repertoire of communicative sounds, postures, and facial expressions, each used as signals of the chimp's mood, location, or desires. As with humans, some signals are intentional, such as a victim's outstretched arm to an aggressor following an attack—signaling that he would like to

above: *Two chimpanzees utter the characteristic pant hoot to communicate with others in the distance. Each individual's pant hoot is unique, and researchers believe the call is used to announce one's presence or solicit a response from companions in the vicinity.*

make peace. Some signals, on the other hand, are unintentional, such as a subordinate's emotionally charged fear grimace, with teeth exposed and lips pulled back. Although the fear grimace resembles a human smile, and is often used in the entertainment industry to portray "happy" chimpanzees, the emotion that induces such an expression is anything but happy. A cheerful and content chimpanzee has a relaxed face with drooping bottom lip. And when engaged in rigorous play the chimpanzee opens his mouth wide, bearing his bottom teeth and covering the top with his upper lip, an expression referred to by primatologists as the play face. Other common chimpanzee facial expressions include the lip-puckering pout and the stern, compressed lips of an aggressive male beginning his display.

Several times each day the African rain forest reverberates with the long, piercing cry of the chimpanzee pant hoot. It begins with a building series of long, boisterous pants and ends with short staccato hoots. The purpose is to communicate with chimpanzees out of sight, across the valley or in a nearby tree. The messages are varied, from "we've found a fruiting tree," to "I'm over here, where is every-

below: *When two bonobos bicker, they often alternate vocalizations as if engaged in a war of words. Note the humanlike bipedal stance of the bonobo on the right, a posture more common among bonobos than chimpanzees because of their elongated legs and straighter backs.*

body else?" Sometimes a listening party responds to the calls by joining the callers in a nearby tree. Sometimes the recipients call back to a lone caller, and he goes to join them. Just as with people, each chimpanzee has his own voice and can be recognized by his pant hoot alone.

Other chimpanzee vocal communications include food grunts when snacking on a favorite food; waa-barks emitted by an angry or threatening individual; full-blown screams during temper tantrums; and breathy laughter during play.

Gestures and postures play important roles in chimpanzee communication, many of which are startlingly similar to our own—open-mouth kissing to resolve conflicts, embracing to welcome a returning family member, touching to soothe a frightened companion, tickling to amuse an antsy youngster, and, in less positive conditions, hitting the arm into the air or in the direction of an adversary to express irritation or to threaten.

Communication is yet another area in which bonobos and common chimpanzees divide. Although the two species do share many facial expressions, bonobos possess several unique vocalizations. While the chimpanzee emits a long, building pant hoot, a bonobo's call in similar situations consists of shrill, yapping "whoops." Bonobos also appear to "chatter" more than chimpanzees in social situations, continuously exchanging peeps and barks. And when two males engage in a confrontation, they alternate vocalizations like dueling banjos. A startling nonverbal communication recently discovered in bonobos is trail marking—leaving visual signs such as leaves or twigs that other members of the party can follow easily.

Closing the Divide

A key element in the communication methods discovered in great apes is their relation to the animals' emotions—fear, anger, joy, and grief. The very notion that great apes, or any other animals, possess emotions is not terribly surprising or earth-shattering today, yet it was not long ago that the mere suggestion was deemed "disgusting" and dismissed as "anthropomorphic." When Wolfgang Köhler and Robert Yerkes published the results of their studies stating that the chimpanzees used reasoning and imagination to solve some of the problems, much of the scientific community was outraged. Despite the obvious similarities among great apes and humans in brain structure, central nervous system, and even behavior, many early scientists believed that animals simply do not have minds; that they operate through inborn responses to stimuli.

As we continue behavioral studies of great apes in the wild, the question has long ceased to be "do they have minds?" We now ask "what is it that makes us different from them? Where do we draw the line?"

Anyone who has ever goaded his dog into "admiring" himself in a mirror knows that dogs simply don't understand mirrors. They don't, as scientists put it, have a concept of self-awareness or self-recognition. In fact, it is believed that only two types of beings in the world are able to look in a mirror and recognize themselves: humans and great apes. Scientists discovered this when they put a small colored dot on an ape's forehead and then placed him in front of his own reflection. Unlike most nonhuman primates, who continually threaten the sudden "intruder" in the mirror, the ape begins to realize that he is looking at himself—evidenced by his eventual attempt to remove the strange spot on his forehead by rubbing and scratching at it. They also contort their faces, inspect areas of their body not easily seen, or place objects on their heads to see how they look while gazing at themselves in a mirror.

A concept or awareness of self leads to the presumed ability to distinguish oneself from other beings: while you are here in this situation, he is over there experiencing something entirely different. This ability to view a situation from someone else's perspective is yet another cognitive skill great apes share with humans—a capability that leads to empathy, altruism, and also deceit.

One of Goodall's most beloved tales of altruism among chimpanzees involves a young female and her infant brother. As the two walked through the forest, sister in front and brother lagging behind, a large snake appeared in the middle of the trail. Pom, the sister, uttered an alarm and quickly climbed a nearby tree. But young brother Prof did not see the snake and continued walking straight into the snake's path. Pom could have stayed in the sanctuary of the tree. But seeing that her brother was in great danger, she swept down from the branches—hair bristling and fear grimace in place—and pulled her brother into her arms and into safety.

Primatologist Frans de Waal describes an act of empathy among the bonobos of the San Diego Zoo, where a six-foot (2m) deep dry moat surrounds the outdoor enclosure. A chain hangs down from the main enclosure to the moat, allowing the bonobos to climb back and forth at their leisure. But as de Waal witnessed many times, one bonobo in particular enjoyed pulling the chain up after another had descended, thereby preventing the bonobo in the moat from returning to the main enclosure. This is not empathy; this is play. But when another bonobo, Loretta, rushed to the aid of the individual in the moat by returning the chain to its position, she clearly understood the predicament of the bonobo in the moat and made an effort to help. As de Waal describes it in *Peacemaking Among Primates*, "[both bonobos] knew what purpose the chain served for someone at the bottom of the moat and acted accordingly; the one by teasing; the other by assisting the dependent party."

Caretakers of captive great apes around the world could fill volumes with stories of the deceptive nature of their intelligent companions—from orangutans pretending not to notice dropped keys until the keeper has left the room (and then using the keys to escape) to signing chimpanzees who deny making a mess or taking a toy when their trainer knows otherwise. Once again, the bonobos at the San Diego Zoo provide an excellent example. When a tiny bonobo named Laura was encouraged by her caretaker in the zoo nursery to finish her food, she immediately obliged. Only later did the caretaker realize how the young bonobo had "finished" the meal so quickly—Laura had stuffed the entire contents of her food dish into her diapers. (Diapers, incidentally, are worn only when the infants are in the nursery because their mothers are unable or unwilling to care for them.) The act of deception requires sophisticated reasoning and a certain amount of imagination. The deceiver must have an idea of how much the other party knows and how far they can go to avoid getting caught in the lie.

The great apes' capacity to experience fear, joy, grief, and anger; their ability to feel empathy for others and act altruistically in the face of danger; and even their tendency to deceive—all point to their incredibly complex minds and their singular status as humans' closest genetic link.

right: *Great apes, like this young bonobo from the San Diego Zoo, exhibit daily their keen intelligence, capacity for emotions, and similarity to humans in a number of ways.*

frans de waal, b. 1948

In a special division of the Yerkes Regional Primate Research Center in Atlanta, Georgia, the study of humans and primate relations focuses not on the missing link, but on the ever active, beguiling, and fascinating Living Links: bonobos, chimpanzees, gorillas, and orangutans. Through noninvasive research, the Living Links Center peers into the social life, ecology, cognition, and genetics of apes in an attempt to reconstruct human evolution, identify the differences between humans and apes, and promote the well-being of the great apes in captivity and in the wild.

At the helm of this unique research center is Dr. Frans de Waal, a Dutch zoologist and ethologist whose own research among the higher primates has opened the eyes and minds of many to the complex social lives of the great apes. De Waal obtained his Ph.D. in biology from the University of Utrecht in 1977, with a dissertation on aggressive behavior and alliance formation in macaques. This research interest was to stay with him as he pondered the 1963 thesis of Konrad Lorenz, the so-called father of ethology, whose message was that humans possess a violent instinct that is virtually impossible to control. De Waal resolved to investigate the phenomena of aggression, conflict resolution, and peacemaking by studying a group of chimpanzees at the Arnhem Zoo in the Netherlands.

He began his research in earnest in 1975 after witnessing an impressive charging display and attack on a female in the group by the dominant male. The attack was followed by an eerie silence, a round of pant hoots, and one male pounding on metal drums, while the dominant male and his female victim kissed and embraced amid the excitement. This, he realized, was an act of reconciliation. And so began a fifteen-year study of the Arnhem chimpanzees, the bonobos of the San Diego Zoo, and two species of macaques (rhesus and stump-tailed). His 1989 book, Peacemaking Among Primates, *which describes his findings and conclusions, received both popular and critical acclaim.*

De Waal later shifted his attention to the proposed core of conflict resolution and peaceful interaction: morality. In his 1996 book, Good Natured: The Origins of Right and Wrong in Humans and Other Animals, *he states, "Given the universality of moral systems, the tendency to develop and enforce them must be an integral part of human nature." He therefore set himself on the search for the "building blocks" of morality in other animals.*

De Waal's study and written conclusions on social aggression and resolution in captive chimpanzees, bonobos, and other primates has been heralded as some of the finest research on nonhuman primates in this generation. He turns a mirror on our species and humbles us with a glimpse into the apes' strikingly similar world. His current work as director of the Living Links Center focuses on food sharing, social reciprocity, and the continued look into the origins of morality and justice in the human race.

chapter v

fighting for survival

Ninety-six hours from the moment you read this sentence, one million more humans will inhabit the earth. And one year from today, fifty thousand more animal species will have disappeared into oblivion, some becoming extinct before we ever knew they existed.

We human beings are literally bulldozing our way across valleys, fields, meadows, and forests, pushing farther into territory that has been home to other species for millenia—territory not big enough to share.

Where do the animals go? When scientists discuss the extinction of a species we imagine an animal suddenly blipping out of existence. One moment it is nesting in a tree, crawling through the bush, or swimming upstream for survival. And the next moment it is gone, taking with it all memory of its ancestors and promise of future generations, without explanation or regret. The truth is not so simple or clear-cut. Extinction has causes: loss of habitat, scarcity of prey, increase in predators. When a species loses its territory, food sources, and protection from new and mounting dangers—including humans—it dies off slowly and painfully, each generation becoming weaker and more vulnerable, until the last animal of its kind breathes its final breath. From Britain's tortoiseshell butterfly, last seen in the wild in 1993, to the tigers of southern China, who now number only around sixty individuals, species the world over are losing ground to the ever-growing human population.

Great apes, the animals so like us, are no exception. Deforestation and commercial logging in Africa and Asia, a growing demand for great ape meat in both rural villages and high-priced Western restaurants, and the ongoing practice of taking young infants from the wild to be sold as pets leave the great apes fighting for survival.

opposite: *The fate of the great apes hangs in balance. Without intervention from conservationalists and concerned citizens, populations of wild apes will continue to shrink.*

Nearly one million chimpanzees once roamed undisturbed in forests along the equatorial belt of Africa. Conservative estimates for chimpanzee populations now hover around 160,000. Bonobos, whose dependence on life in the trees limits their adaptability to other ranges, live only in the remaining forests south of the Zaire River. Fewer than ten thousand wild bonobos remain.

Gorillas face the same fate in the African forests. Fewer than one hundred thousand western lowland gorillas now live in Western Africa. Ten thousand eastern

Estimated range 10,000 years ago

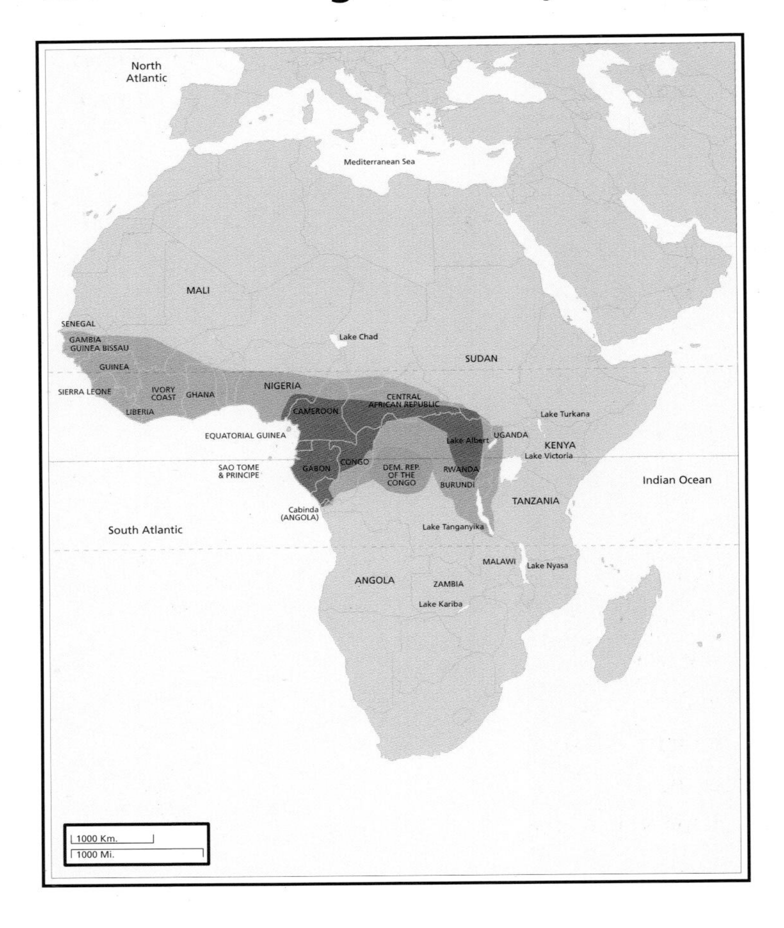

Current range

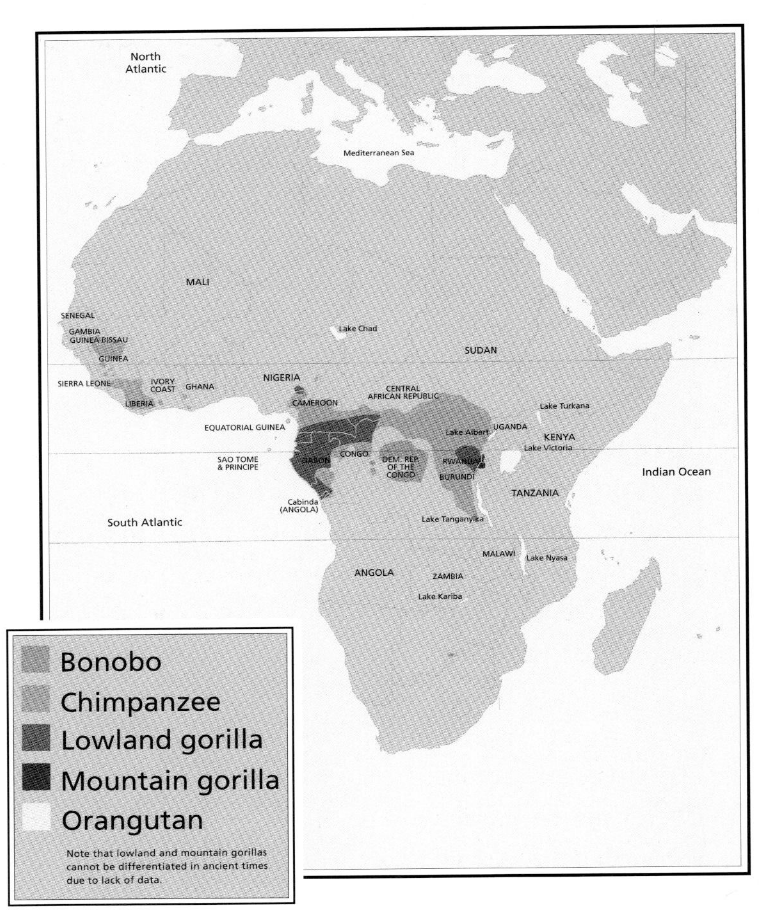

Estimated range 10,000 years ago

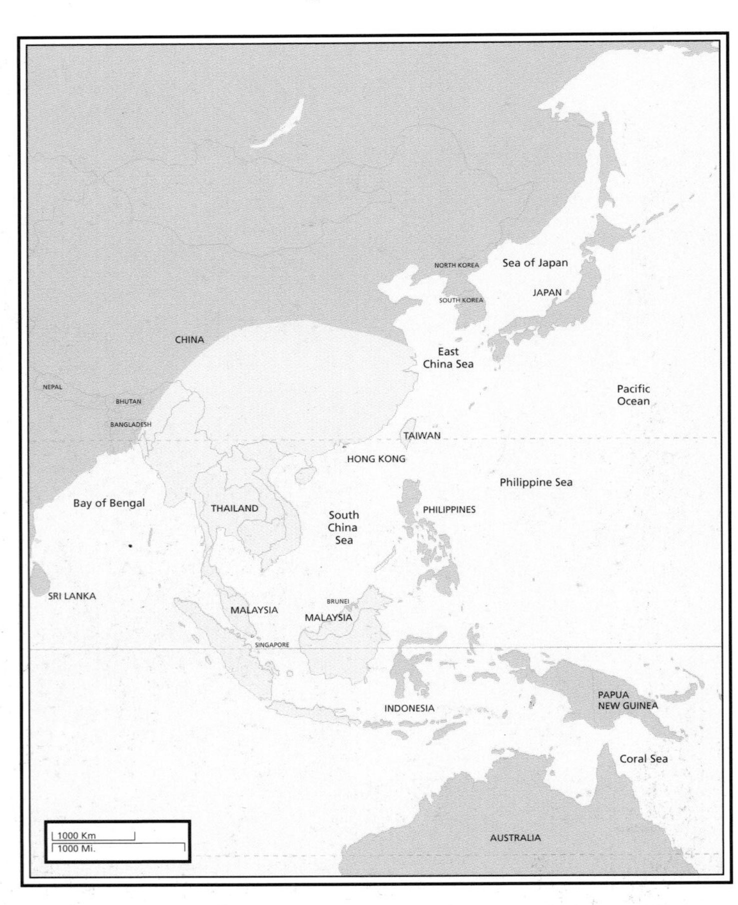

Current range

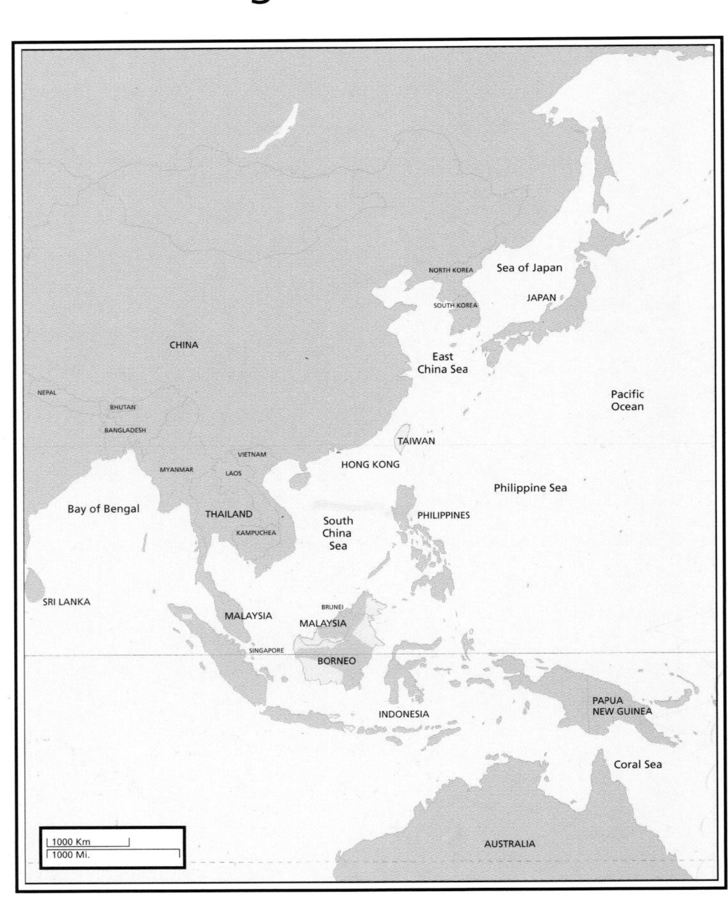

lowland gorillas roam the dwindling forests of eastern Congo. And the mountain gorillas, the subspecies Dian Fossey gave her life to protect, have now declined to the perilously low population of six hundred fifty.

Finally, the orangutan—the shy red ape whose very threat to survival brought it to the attention of the unknowing world—inches its way closer to extinction each day. Experts believe hundreds of thousands of orangutans once lived throughout southern China and through southeast Asia. Now twenty-five thousand cling to the scattered forests of Borneo and Sumatra. At least one thousand orangutans died in 1997 alone during the devastating forest fires that tore through Indonesia. Some scientists predict that they will be extinct in the wild within twenty years.

opposite: *These maps portray the degree to which great ape ranges have contracted over the past ten thousand years.*

below: *Slash-and-burn agriculture and other invasive land uses have taken a heavy toll on great ape habitats.*

Nowhere to Go

Indigenous peoples of Africa and Asia have lived on the bounty of the forests for centuries, taking from the sacred forests only what they needed, and always with respect for the animals who gave their lives to ensure the people's survival. These people were a part of the rich cycle of life and death—just like the animals who shared their forest homes. It is the exponential growth of the human population that has drastically upset this fragile balance. Many of those who lived in the forests have been forced to move to villages, to develop a more "modern" lifestyle of subsistence farming. Each small farm replaces forest, savanna, or woodland, destroying the precious biodiversity of the region and depleting the soil to the degree that the land will eventually stop supporting crops.

above: *Widespread commercial logging and devastating fires rip through the Indonesian forests, destroying precious orangutan habitat. Solutions for the apes' long-term survival must start with the preservation of their forest home.*

As people encroach further on the forest, the resources of the land offer greater allure—wood for coal, bamboo for fences, duiker for food, great apes for sale.

To address these challenges, the conservation community realized it must provide alternate means of survival for those who would otherwise destroy the earth's rain forests and their inhabitants. But just as environmental education and community-based conservation projects began to spring up across Asia and Africa, a bigger, more powerful threat loomed: commercial logging.

Where once a few men bushwhacked their way through their forest to hunt for meat or a small number of women followed a narrow, well-worn trail to gather firewood, commercial logging operations now bulldoze or burn their way into once untouched forests. Large logging roads replace dense undergrowth where young gorillas once played. Idling trucks echo the distant calls of frightened chimpanzees whose territory is dwindling. And forest fire smoke billows through the trees, forcing arboreal orangutans to search in vain for shelter on the burning forest floor. Throughout Indonesia, it is believed that the fires were started by humans, rather than by natural causes such as lightning, though controversy surrounds the naming of the guilty parties.

The 1997 fires that devastated Indonesia burned five million acres of forest, contributing to researchers' estimates that within the past decade 80 percent of the orangutan's habitat has been destroyed by commercial logging and land clearance.

left: *The poaching of young orangutans rose when the fictional family on a popular Taiwanese television show adopted an orang as a family pet. Although the demand in Taiwan has slowed, orangutan mothers continue to be killed and infants sold to fuel the pet trade in other parts of Asia.*

right: *An adult male Bornean orangutan camouflages himself among his long red locks that flow from back, shoulders, and arms. Because orangutans spend their lives in the trees, they rarely see or are seen by the villagers who live on the margins of the forest. When the forests are destroyed by loggers, the orangs must flee to the villages for shelter and food, where they are usually shot.*

When the orangutans lose their land, they are forced to cross into the territory of humans—raiding gardens and palm oil plantations to find food to survive, and losing their lives to humans who will not, and should not, compete with the great apes for their own family's survival.

Recent logging in the central African country of Gabon has even initiated territorial warfare among two troops of chimpanzees, wiping out nearly 80 percent of each community's population. Chimpanzees are ardent defenders of their turf; advancing logging operations and destruction of habitat forces neighboring communities into each other's territories, setting off violent and deadly battles to retain the remaining forests.

Gorillas, sometimes fighting for the same forests as chimpanzees and bonobos, do not face territorial conflicts with neighboring troops. But their disappearing habitat threatens their existence just the same. Where five hundred years ago the mountain gorilla habitat stretched uninterrupted through Rwanda, Congo, and Uganda, it is now divided by deforestation and logging, creating two shrinking islands of mountain gorilla populations in a sea of human developments.

Even when logging is selective—an attempt to manage rather than destroy a forested area—the rich diversity of the forest cannot emerge unscathed. Large trees damage

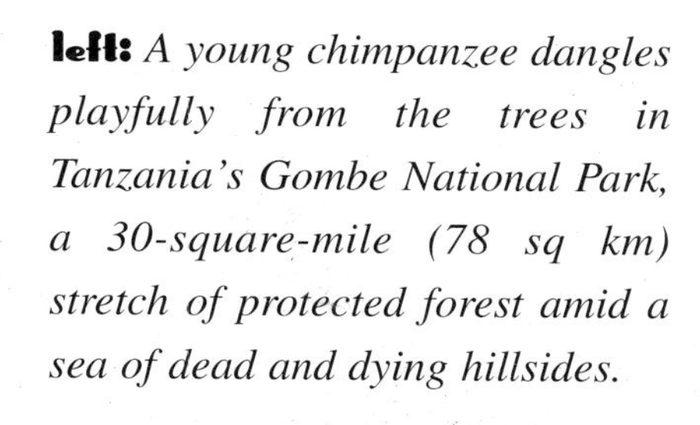
left: *A young chimpanzee dangles playfully from the trees in Tanzania's Gombe National Park, a 30-square-mile (78 sq km) stretch of protected forest amid a sea of dead and dying hillsides.*

small trees when they are felled. The biodiversity of the forest suffers when only certain species of trees are harvested. Fish die when streambeds are used for roads. And roads provide poachers with easy access to previously inaccessible hunting grounds.

The Appetite for Bushmeat

Small-scale hunting is a traditional way of life in Western and Central Africa. But the easy access deep into forests provided by logging roads has turned subsistence hunting for the purpose of feeding one's family into a commercial business that is threatening to wipe out the African great apes. Primatologists studying the effects of the bushmeat trade estimate that in one year hunters illegally kill and butcher more than two thousand gorillas and four thousand chimpanzees—ten times the number of chimpanzees who live in Goodall's research area of Gombe Stream. Some estimate that three times as many great apes are eaten each year than live in zoos and laboratories in North America.

The process is brutal and remorseless. After killing the apes with shotguns, the hunters hack their bodies into pieces with machetes, smoke the pieces to preserve the meat, then transport the game on logging trucks into nearby villages and urban centers, where it is sold as a delicacy. Gorilla hands and heads are sold as fetishes for their reputed mystical properties.

Even in areas where local religious beliefs prohibit the hunting and killing of great apes, the snares set for other animals (bushpig, duiker, and so on) do not discriminate. Many chimpanzees, gorillas, and bonobos lose hands and feet in snares meant for their forest neighbors, and many soon die from infection or loss of blood.

Although large-scale hunting of orangutans for meat has not been a threat for hundreds of years, indigenous people feeling the effects of food shortage caused by fires and logging are forced farther into the forests, where the slow-moving orangutans become ready targets.

Capture for Pets

Perhaps the most tragic consequence of great ape poaching is the accompanying capture of live infants. Youngsters—usually under the age of four, still nursing and extremely dependent on their mothers—are the tiny bonuses in a hunter's day. In search of the meat of adult apes, hunters kill the mother and other nearby party members and tear the infant from the mother's body to sell in the village as a novelty or pet—their small bodies of no use as a source of protein. At least ten chimpanzees and gorillas die for every infant captured. Often the infant is thrown in a burlap bag atop his mother's mutilated corpse—an unimaginable horror for such an intelligent and sensitive being.

right: *Less than one week previously this infant chimpanzee lay nestled in the comfort of his mother's arms in the rich African rain forest. Poachers killed his mother for meat and captured him to sell as a pet. Still in shock after seeing his mother brutally killed, and fed only bananas and scraps by his captors, he'll likely die before he reaches the age of one.*

Where the traumatized infants go next depends solely on the entrepreneurial savvy of the hunters. Sometimes infants are merely taken home by the hunter as a pet or plaything for his children. Other times they are carted to villages to sell at local markets.

Because infant apes rely so heavily on their mothers for nutritional and emotional support, most kidnapping victims die of illness and trauma within the first few days after their capture. Their new caretakers throw them bananas or other solid foods they are unprepared to eat, and wonder why the food goes untouched. They lock the infants alone in small crates or tie them to trees and question why they cower, instead of dancing playfully for the children's entertainment.

When great apes kept as pets do survive, they may at first be pampered as the newest member of a growing family. Their resemblance to humans places them in the role of spoiled child, where they are dressed in clothes, allowed to eat in the house, and given free range of the family dwellings. But young great apes do not stay docile for long. In the wild, they would be exploring their surroundings, experimenting with aggressive play, and challenging the authority of their mothers and other elders on their long road to adulthood. In the human household, their new aggression and unmanageability leads to a life in chains. When great apes begin to bite, rip curtains, raid refrigerators, and frighten children, they are ban-

Below: *Efforts to reintroduce rescued chimpanzee infants to the wild have been largely unsuccessful because the youngsters have not yet learned the necessary survival and social skills. Those who cannot be returned to the wild live in special preserves, such as this one in Zambia.*

right: *Great ape infants like this young orangutan appeal to ignorant buyers because of their similarities—in appearance and behavior—to human babies. Yet they do not stay young, small, and huggable for long.*

ished to pens in small dark corners of the house or chained, like dogs, in the yard. Sometimes they are killed.

In Africa, most orphaned gorillas, chimpanzees, and bonobos go home with sympathetic but ignorant tourists or Western expatriates, who buy them out of pity. The buyers fail to realize that by paying money for one small, ailing infant they are contributing to and encouraging the trade, leading poachers back into the forest to steal yet another infant from his forest home. Before the great apes were declared endangered, it was common practice to capture infants from the forest for export to North America and Europe for circuses, zoos, the exotic pet trade, and the growing medical research industry. Although it is now illegal to export endangered species internationally, many young apes are discovered by airport officials in suitcases and crates, bound for horrors unknown.

In Asia during the mid- to late 1980s, a popular Taiwanese television show featuring a young orangutan as a family pet led to the smuggling of nearly one thousand orangutans to the island of Taiwan and into the homes of ignorant but wealthy families who sought to emulate their favorite program. When the families realized that the solitary, intelligent, and powerful orangutans did not make good pets, the young apes were abandoned. Although the demand for orangs in Taiwan has subsided, hundreds of infant orangs continue to be stolen from the forests for sale in Indonesia and Malaysia to anyone with the money to buy.

left: *Sanctuaries like this one in Congo become easy drop-off sites for those people who lose interest in their chimpanzee pets. Most sanctuaries not only provide a home to orphaned great apes but work with various governments to enforce laws against poaching and educate the local people on the long-term benefits of conserving their forests and the animals who live there.*

Solutions

The first step in solving a problem is to recognize that the problem exists. For the great apes, this step occurred when they received protection under both the United States Fish and Wildlife Service and the Convention on International Trade in Endangered Species of Wild Fauna and Flora (CITES).

When the U.S. Endangered Species Act passed in 1969, it provided protection to species worldwide—not just in the United States—by preventing import of endangered animals into the country and prohibiting their subsequent sale. This single act drastically curtailed the capture of infant great apes for U.S. circuses, zoos, and laboratories.

The gorilla and orangutan were listed under the U.S. Endangered Species Act in 1970; chimpanzees and bonobos in 1976. All four species have been elevated from the less restrictive classification of "Threatened" to "Endangered" within the past twenty years—with one exception. Captive chimpanzees held in the United States remain listed as threatened, a political move that allows for the continued sale and transport of chimpanzees across state lines for specific purposes, including medical research.

In 1973 government officials from around the world signed the international treaty known as CITES to prevent the overexploitation of wildlife caused by international trade. The purpose of CITES is to ban the international trade of an agreed-

above: *Terraced farming creeps closer and closer to the boundaries of Virunga National Park, home to gorillas and chimpanzees.*

upon list of endangered species—both flora and fauna—and regulates and monitors the trade in species that are at risk of overexploitation. Species listed as Appendix I face the greatest risk of extinction. With some exceptions, species listed as Appendix I are banned from international commercial trade. Gorillas and orangutans were listed as Appendix I in 1975; chimpanzees and bonobos in 1977.

Although CITES has a membership of more than 140 countries, all vowing to protect the world's endangered species, compliance in some developing countries is often minimal. In short, the recognition that our closest relatives are in danger of extinction is an important step, but solutions to the causes of their fragile status is what will turn the global trend around.

The greatest threats to great ape survival—deforestation, bushmeat hunting, and the pet trade—intertwine. Logging destroys habitat and creates better access for bushmeat hunters, which leads to more butchered and smoked great ape meat in the markets and more orphaned infants languishing on chains in the tropical sun. The solutions to these problems, therefore, begin with the protection of the forests.

Conservation organizations are urging governments to curtail or terminate commercial logging of the fragile forests. Income from the commercial logging industry can be replaced by establishing the forests as national parks, where nature-hungry ecotourists will pay upwards of $100 an hour to catch a glimpse of the majestic and endlessly fascinating great apes. In the Virunga Mountains in Rwanda, Uganda, and the Democratic Republic of Congo, for example, a gorilla-trekking ecotourism project started by the International Gorilla Conservation Program in 1980 helped the mountain gorilla population in one region grow from only 250 individuals to 320. And according to the World Watch Institute, the Rwandans were so dedicated to the success of the program that during the country's bloody civil war in 1994, both sides agreed that the mountain gorillas should be protected from the fighting.

above: *Guards in Virunga National Park patrol the park boundaries, searching for signs of poaching, illegal logging, and snares. Many guards in national parks throughout Africa are former poachers themselves, their keen tracking skills and familiarity with the forests put to better use by conservation organizations who provide a stable salary and ongoing environmental education.*

Perhaps the best example of a national park established for the benefit of the great apes is Gombe National Park in northern Tanzania, the site of Jane Goodall's ongoing study of chimpanzees. When Goodall arrived in Tanzania in 1960, she set up camp in what was then Gombe Stream Reserve. In 1968 the Tanzanian government declared it a national park, further emphasizing their commitment to Dr. Goodall's research and the protection of her research subjects. Today, as you approach the park from Lake Tanganyika, the ten-mile (16km) stretch of Gombe stands out as the sole surviving forest surrounded by miles of terraced, cultivated, and deforested farmland where lush green forests once stood. Little do the chimpanzees of Gombe realize how fortunate they are to have been visited by the determined young woman from England.

A successful transformation from hunting ground to tourist destination requires the interest and commitment not only of the government but of the local people—those who have grown accustomed to exploiting rather than protecting the nearby forests. It is unproductive for outside organizations to sweep into a country and dictate to the local people that the lives of the animals are more important than their own livelihood and their ability to provide for their families. Dedicated conservationists must collaborate with the local people, who face questions of survival themselves, helping to establish the long-term advantages of protecting wildlife and natural resources and working to develop alternate means of subsistence.

In 1994 the Jane Goodall Institute expanded its conservation efforts in Tanzania to address the mounting deforestation that pressures Gombe National Park. The Lake Tanganyika Catchment Reforestation and Education project (TACARE) in the Kigoma village region, just ten miles (16km) south of Gombe, provides villagers with training and supplies to create their own nurseries of fruit trees, hardwood trees, and vegetables, both for personal use and to sell at local

above: *A resident at Sweetwaters Chimpanzee Sanctuary in Kenya takes a break from rambunctious rough-and-tumble play. Although chimpanzees are not indigenous to Kenya, the sanctuary provides 200 acres (81ha) of riverine woodland to orphaned chimpanzees from surrounding countries, including Burundi, Rwanda, and the former Zaire.*

markets. By providing the people with alternatives to raiding the forest trees and establishing agricultural methods other than slash-and-burn farming, TACARE helps protect the forest while helping the people improve their standard of living. TACARE also integrates the Jane Goodall Institute Roots & Shoots component into their training; this environmental and humanitarian education program for children and adults encourages people to show care and concern for their environment, animals, and their own communities. And because women in this region of Tanzania are the family providers, future phases of TACARE will include women's health and family planning, educational scholarships for young women, and a microcredit loan program to help women start their own small businesses.

In other projects, former hunters and loggers can become well-trained and respected trackers, guiding visitors through the forests to view the gorillas, chimpanzees, or bonobos, or they may choose to work as armed eco-guards, canvassing the forest for poachers.

The Bonobo Protection Fund (BPF), established in 1990, operates conservation research in the bonobo habitat of Democratic Republic of Congo. Although the country has experienced political turmoil that has frightened many conservation organizations away, BPF recognizes that the bonobos have little time to lose. Although the hunting of bonobos is illegal in the Congo, the laws are not enforced. BPF distributes educational materials to the people who live on the margins of bonobo habitat and employs many local people to maintain the bonobo research facilities and assist with building and maintenance of forest research trails. In return, they receive basic living staples, which reduces their need to hunt and sell meat for survival. The next steps for BPF include setting up a conservation and education research center in Kinshasa, the capital city, to be staffed by Congolese workers.

In Uganda, the International Gorilla Conservation Program is active in the Mgahinga Gorilla National Park and Bwindi Impenetrable National Park, creating buffer zones to protect against encroaching cultivation and further develop successful gorilla-trekking projects that bring in needed revenue and respect the space and solitude of the resident gorillas.

Also in Uganda, in the Kibale Forest, the Jane Goodall Institute has hired a community warden to canvass villages surrounding the park in order to learn about local attitudes regarding the protected areas and devise ways to educate the people on the need for conservation.

Like-minded conservation organizations are also joining forces to address the problem of deforestation. The African Forest Action Network is a coalition of approximately thirty nonprofit organizations in West and Central Africa interested in conservation and sustainable use of forest resources. The Network disseminates information about deforestation issues in the region and coordinates lobbying efforts. The Global Forest Policy Project, founded in 1992, works to reduce the rate of deforestation, expand the protection of important forest areas, and implement sustainable management of the world's forests.

In 1996 the Biosynergy Institute in Los Angeles developed the Bushmeat Project to raise awareness about and provide solutions to the growing problem of great apes ending up in meat markets, especially in Africa. The project encourages logging companies to provide domestic meat for their workers and families, conduct courses on wildlife laws, create teams of villagers to patrol the forests for illegal or unsafe hunting and logging practices, and to destroy bridges and roads when the logging operations are complete. They also hope to create alternatives to bushmeat by establishing village farms that raise domestic animals for market. Perhaps most important, the Bushmeat Project is working with African governments to encourage the implementation of national and regional bushmeat policies and practices. The World Society for the Protection of Animals is also active in the fight against the African bushmeat trade.

Although the threats to the orangutans are similar to those of the African apes, the culture requires different solutions. Laws against the owning, killing, or harming of orangutans were passed in Indonesia in 1931, in Sarawak in 1958, and Sabah in 1963. But aggressive logging in the orangutan's habitat has forced conservation organizations to step up their efforts to save the species from annihilation.

The Orangutan Foundation International, established and directed by Dr. Birute Galdikas, operates a rain forest restoration program in which local women are hired to plant various species in degraded areas outside Tanjung Puting

National Park, site of Galdikas's research base, Camp Leakey. Flora planted included light-tolerant canopy species and shade-tolerant understory species. OFI's Orangutan Research and Conservation Project supports the ongoing behavioral study of wild orangutans and encourages the conservation of the red ape and its rain forest habitat. In cooperation with the Indonesian government, ORCP conducts patrols of the park and assists government officials in identifying and curtailing illegal poaching and logging operations. The Orangutan Foundation International also sponsors ecotourism trips to view wild and ex-captive orangutans living in Tanjung Puting Park.

The devastating Indonesia forest fires of 1997 spurred many other organizations to action as well. The World Wide Fund for Nature, the U.S. Pittsburgh Zoo, and the Wildlife Conservation Society, among others, are working to determine the severity of the loss of orangutan lives and habitat during the fires and protect the remaining individuals in the devastated areas. The Wildlife Conservation Society's Field Vet Program has even relocated some wild orangutans from the destroyed area to new forest areas uninhabited by humans.

Whether in Asia or Africa, the rescue of the wild great apes begins with encouraging local governments to enforce the laws and providing the people with an alternate source of income and survival, all while engendering appreciation and respect for the fragile forests and the great apes who live there.

A Place to Call Home

But what of the surviving victims of poaching, the young orphans who face an unnatural life as a pet? Like humans, the great apes' childhoods are long. In the wild, they grow up under the watchful eye of their mothers, learning the skills of survival as well as the social graces they will need to succeed as an adult. Great apes deprived of this maternal care are not prepared for life in the forest.

Attempts to return orphaned chimpanzees, gorillas, and bonobos to the wild have failed. We cannot shoo them into forests to survive on their own because they rely on social bonds, community support, and tutelage from their mothers for survival. We cannot thrust them into an existing community because they will likely be killed as intruders. When orphans young enough to be adopted by an elder are eased into an existing community by well-meaning conservationists, the youngsters generally return to the village or camp from which they came for the easy handout of food.

To enforce the laws against poaching and trade, governments are encouraged to confiscate the infants before any sales can be made and to fine the poachers to discourage them from trying again. Various organizations throughout Africa provide the young orphans with care, nurturing, and companionship in protected areas, where they can once again forage, climb trees, develop social bonds, and feel the grass on their feet. For the rescued chimpanzees, bonobos, and gorillas who can't be returned to the wild, these sanctuaries become the apes' lifelong homes. Although the poachers forever robbed them of true freedom, the sanctuaries staffed by caring professionals save them from desolate lives as ex-pets and storefront sideshows.

The solitary nature of orangutans makes them prime candidates for potential reintroduction because they needn't fear alienation or territorial violence from others of their kind. Not long after Birute Galdikas began her study of the wild orangutans in Borneo, she became aware of the vast number of orphaned orangs who had been taken from their mothers and put up for sale as pets. Camp Leakey soon became home to ex-captive orphans who had been confiscated from poachers by the government. Galdikas cared for the youngsters during their infancy and slowly introduced them into the reserve around the camp, where they could once again roam free. The growing rehabilitation program has now moved to another area of the park, and a new Orangutan Care Center and quarantine facility in Central Kalimantan is under way.

Also working to rescue and rehabilitate captive orangutans is the Wanariset Orangutan Reintroduction Centre, a joint venture between the Indonesian Ministry of Forestry and the Netherlands Ministry of Education. The Centre, situated on 8,655 acres (3,504 ha) of forestland, takes in orphaned orangutans from families, traders, and government officials and returns them to the jungles of East Kalimantan. Since the Centre's opening in 1991, two hundred orangutans have landed on their doorstep and 110 have been reintroduced into the wild.

below: *A young Bornean orangutan awaits transport to a reintroduction facility after being confiscated from poachers.*

closing statement

our own backyard

The thinking man must oppose all cruel customs, no matter how deeply rooted in tradition or surrounded by a halo. We need a boundless ethic which will include the animals.

—Dr. Albert Schweitzer

Louis Leakey sent three women into the rain forests of Africa and Asia in search of living links to our human past. By understanding the great apes, he reasoned, we could better understand ourselves.

The three women Leakey sponsored in these studies, and countless others who have followed in their footsteps, soon realized that the study of great apes need not be justified by the search for early man. Their lives in the wild, their relationships, behaviors, struggles, and triumphs are worthwhile because the gorillas, orangutans, chimpanzees, and bonobos are fascinating, complex creatures in and of themselves.

And so it must be with their fight for survival. Regardless of one's views on evolution, the missing link, or early humans, great apes are, without doubt, worthy of our protection. That they exhibit emotions, personalities, and behaviors similar to our own—and can communicate with us on our own terms—opens a narrow window into their complex lives. We may never know with certainty the message in the lonesome long call of the male orangutan, the purpose of the stalk of grass held between the lips of a fighting silverback gorilla, or the depth of love and understanding between a mother chimpanzee and her adult son. Yet we will certainly never get the chance to unravel these mysteries if the great ape populations continue to decline.

Declaring the great apes endangered in the wild helps raise awareness of their fragile status and declining habitats. Yet with one breath we lament their possible disappearance and strive to convince indigenous people of the animals' uniqueness, and with the other we laugh heartily at orangutans dressed in clothes for television sitcoms or chimpanzees trained to perform in circuses and movies.

It is largely the more developed nations who have imported great ape infants from their native lands, bred them for our own entertainment, then banished them to lonely cages or research laboratories when they are no longer cute and controllable. It is the residents of these nations, therefore, who can take an active part in the fight for the great apes' survival. We must examine the indignities we've thrust upon these amazing creatures in captivity—the pain, isolation, and humiliation they have suffered in the name of science and entertainment. The intelligent, noble beings we fight to save in the forests of Africa and Asia are merely the more fortunate cousins of those who languish in the zoos and circuses in our own backyard.

The great apes born and bred in captivity will never know the pride and purity of their wild relations—and those in the wild will likewise have no knowledge of captivity, loneliness, degradation, or boredom—yet they all deserve the same respect, the same fight, and the same hope for the future. While we push for alternatives to bushmeat, fight for enforcement of poaching laws, and urge commercial logging companies to retreat from virgin forests, we can also turn our backs on circuses and roadside zoos that exploit all animals, boycott films and television shows that feature performing apes, and urge research companies to seek alternatives to the use of chimpanzees as living, breathing test tubes. For only when we begin to bestow upon all great apes the dignity they have earned in the wild will we truly save these magnificent beings from extinction.

left: *A gorilla infant clings to his imposing silverback father for protection.*

organizations

Great Ape Conservation and Study

Bonobo Protection and Conservation Fund
Georgia State University Foundation
University Plaza
Atlanta, GA 30303
404-244-5825

Bushmeat Project
Biosynergy Institute
P.O. Box 488
Hermosa Beach, CA 90254
310-379-1470

Dian Fossey Gorilla Fund
800 Cherokee Avenue SE
Atlanta, GA 30315-9984
404-624-5881

Friends of Washoe
Central Washington University
Nicholson Blvd and D Street
Ellensburg, WA 98926-7573
509-963-2214

International Primate Protection League
P.O. Box 766
Summerville, SC 29484
803-871-2280

The Jane Goodall Institute—USA
P.O. Box 14890
Silver Spring, MD 20911
800-592-JANE
www.janegoodall.org

The Jane Goodall Institute—UK
15 Clarendon Park
Lymington, Hants
England SO41 8AX

Orangutan Foundation International
822 S. Wellesley Ave.
Los Angeles, CA 90049
800-Orangutan

Orangutan Foundation
7 Kent Terrace
London, England NW1 4RP
+44 171 724-2912

Rainforest Action Network
221 Pine St., Suite 500
San Francisco, CA 94104

Wildlife Conservation

Defenders of Wildlife
1101 14th Street, NW #1400
Washington, D.C. 20005
202-682-9400
www.defenders.org

Earthtrust
25 Kaneohe Bay Drive, Suite 205
Kailua, HI 96734
808-254-2866
www.earthtrust.org

International Wildlife Coalition—United States
70 East Falmouth Highway
East Falmouth, MA 02536
508-548-8328
www.iwc.org

International Wildlife Coalition—Canada
PO Box 461
Port Credit Postal Station
Mississuaga, ON L5G4MI
905-279.2043

International Wildlife Education and Conservation (IWEC)
237 Hill Street
Santa Monica, CA 90405
310-392-6257
www.iwec.org

National Wildlife Federation
8925 Leesburg Pike
Vienna, VA 22184
703-790-4000
www.nwf.org

The Wildlife Society
5410 Grosvenor Lane
Suite 200
Bethesda, MD 20814-2197
301-897-9770
www.wildlife.org

World Wildlife Fund
1250 Twenty-Fourth Street, N.W.
PO Box 96555
Washington, DC 20077-7795
www.wildlife.org

bibliography

de Waal, Frans. *Peacemaking Among Primates.* Cambridge: Harvard University Press, 1989.

de Waal, Frans, and Frans Lanting. *Bonobo: The Forgotten Ape.* Berkely: University of California Press, 1997.

Dreifus, Claudia. "A Conversation with Emily Sue Savage-Rumbaugh; She Talkes to Apes and, According to her, They Talk Back." *The New York Times*, April 14, 1998.

Eimerl, Sarel, and Irven DeVore. *The Primates.* New York: Time Life Books, 1965.

Feldmann, Linda. "D.C.'s Newest Think Tank." *Christian Science Monitor*, December 27, 1995.

Fouts, Roger. *Next of Kin: What Chimpanzees Have Taught Me About Who We Are.* New York: William Morrow and Company, Inc., 1997.

Gilbert, Bil. "New Ideas in the Air at the National Zoo." *Smithsonian,.* June 1996.

Goodall, Jane. *The Chimpanzees of Gombe: Patterns of Behavior.* Cambridge: Harvard University Press, 1986.

Goodall, Jane. *In the Shadow of Man.* Boston: Houghton Mifflin Company, 1971.

Goodall, Jane. *Through a Window: My Thirty Years with the Chimpanzees of Gombe.* Boston: Houghton Mifflin Company, 1990.

Hayes, Harold, T.P. *The Dark Romance of Dian Fossey.* New York: Simon & Schuster, 1990.

Hirschberg, Charles and Robert Allison. "Primal Compassion." *Life,* November 1996.

Johnson, George. "Chimp Talk Debate: Is It Really Language?" *The New York Times*, June 6, 1995. p. C-1.

Kano, T. "The Bonobos' Peacable Kingdom." *Natural History,* November 1990.

Kavanagh, Michael. *A Complete Guide to Monkeys, Apes, and Other Primates.* New York: Viking Press, 1984.

Kohler, Wolfgang. *The Mentality of Apes.* New York: Liveright, 1925.

Leiman, Ashley and Nilofer Ghaffar. "Use, Misuse and Abuse of the Orangutan—Exploitation as a Threat or the Only Real Salvation?" In *The Exploitation of Mammal Populations*, edited by Victoria J. Taylor and Nigel Dunstone. London: Chapman Hill, 1996.

Linden, Eugene. "Can Animals Think?" *Time,* March 2, 1993, p. 54.

McGrew, William C., Linda F. Marchant, and Toshisada Nishida, eds. *Great Ape Societies.* Cambridge: Cambridge University Press, 1996.

Macdonald, David, ed. *All the World's Animals: Primates.* New York. Torstar Books, 1984.

Maple, Terry, and Michael P. Hoof. *Gorilla Behavior.* New York: Van Norstrand Reinhold Co., 1982.

Montgomery, Sy. *Walking With the Great Apes.* Boston: Houghton Mifflin Company, 1991.

Morell, Virginia. *Ancestral Passions: The Leakey Family and the Quest for Humankind's Beginnings.* New York: Simon & Schuster, 1995.

Morris, Ramona and David. *Men and Apes.* New York: McGraw-Hill Book Company, 1966.

Nadler, R., B. Galdikas, L. Sheeran, and Rosen, M. *The Neglected Ape.* New York: Plenum Press, 1995.

Redmond, Ian. *Gorilla.* New York: Alfred A. Knopf, 1995.

Rose, Anthony. "The African Great Ape Bushmeat Crisis." *Pan Africa News*, Autumn 1996.

Schaller, G. "Behavioral Comparisons of the Apes." In *Primate Behavior: Field Studies of Monkeys and Apes*, edited by Irven Devore. New York: Hart, Rinehart and Winston, 1965.

Schreuder, Cindy. "Debate Over Conservation of Mountain Gorilla." *Chicago Tribune*, June 13, 1995.

Small, M.F. and F.B.M. de Waal. "What's Love Got to Do With It: Sex Among Our Closest Relatives is Rather Open Affair." *Discover,* June 1992.

Stevens, William K. "Logging Sets Off an Apparent Chimp War." *The New York Times,* May 13, 1997.

Stewart, Kelly. "International Gorilla Conversation Program." *Gorilla Conservation News,* May 1998.

Tuttle, Russell H. *Apes of the World: Their Social Behavior, Communication, Mentality, and Ecology.* Park Ridge, NJ: Noyes Publications, 1986.

Tuxill, John. "Death in the Family Tree: The Global Decline of Primates." *World Watch,* September/October 1997.

Watts, David P., and Anne Pusey, "Behavior of Juvenile and Adolescent Great Apes," In *Juvenile Primates: Life History, Development and Behavior*, edited by M.E. Pereira and L.A. Fairbanks. New York: Oxford University Press, 1993.

Wrangham, Richard and W.C. McGrew, F. de Waal, and P. Helthe, eds. *Chimpanzee Cultures.* Cambridge: Harvard University Press, 1994.

photo credits

APBL Images: ©Nigel J. Dennis: pp. 78-79; ©Martin Harvey: pp. 8, 11, 27, 40, 41, 48-49, 50, 70-71, 76, 114, 123, 132, 133; ©Anup Shah: p. 97

©Steve Arcella: pp. 13, 14, 122 all

Aurora: ©Robert Caputo: p. 128

©Steve Bloom Images: pp. 22, 62, 65, 77, 86, 93, 115, 139

©Connie Bransilver: pp. 19, 47, 73, 137

Bruce Coleman Inc.: ©K&K Ammann: pp. 2-3, 23, 51, 75, 92, 104-105, 113, 124, 129, 131; ©Wolfgang Bayer: p. 130; ©Tom Brakefield: pp. 30-31, 64; ©Joe McDonald: pp. 68, 120; ©R.A. Mittermeier: p. 33; ©Steve Solum: p. 91; ©Norman Owen Tomalin: pp. 15, 66, 74, 89, 96, 126, 127

Dembinsky Photo Associates: ©Stan Osolinski: p. 102; ©Anup Shah: pp. 39 bottom left, 44, 46, 90, 134

©Michael H. Francis: pp. 29, 58, 59, 80, 82, 118

©John Giustina: pp. 3, 6, 12, 18, 24, 25, 38, 39 top and bottom right, 43, 53, 56-57, 60, 81, 99, 101, 106, 107, 140

Courtesy Jane Goodall Institute: ©Stephen Patch: p. 61

©Michael Nichols: pp. 108, 110

©James Prout: p. 39 top left

Tom Stack & Associates: ©Brian Parker: p. 36; ©Larry Tackett: p. 34; ©Roy Toft: pp. 54, 55, 72, 84-85, 98, 109, 116

Visuals Unlimited: ©M. Long: p. 26

The Wildlife Collection: ©Bob Bennett: pp. 52, 112; ©Michael H. Francis: pp. 45, 125; ©Martin Harvey: pp. 16, 20-21, 69; ©Tim Laman: pp. 17, 94-95

index